FINDING GOOD FARMLAND

How to Evaluate and Acquire Land for Raising Crops and Animals

Ann Larkin Hansen

Storey Publishing

The mission of Storey Publishing is to serve our customers by publishing practical information that encourages personal independence in harmony with the environment.

Edited by Deborah Burns
Art direction by Carolyn Eckert
Text production by Teresa Wiscovitch

Cover illustration by © Michael Austin/Jing and Mike Company
Interior illustrations by © Steve Sanford, except for sketches, pages 50–51, by Deborah Burns

Indexed by Samantha Miller

Storey Publishing
210 MASS MoCA Way
North Adams, MA 01247
www.storey.com

Printed in the United States by McNaughton & Gunn, Inc.
10 9 8 7 6 5 4 3 2 1

Library of Congress Cataloging-in-Publication Data

Hansen, Ann Larkin.
Finding good farmland / by Ann Larkin Hansen.
p. cm.
Includes index.
ISBN 978-1-61212-086-7 (pbk. : alk. paper)
ISBN 978-1-60342-874-3 (ebook)
1. Land use, Rural. 2. Farms, Small. 3. Organic farming. I. Title.
HD256.H36 2013
333.76—dc23

2012045888

CONTENTS

To the Reader

This book assumes that you are looking for rural property to use as your primary residence with the intention of engaging in small-scale organic farming of some type (livestock, dairy, fruits, vegetables, field crops, specialty crops, or a little bit of everything). I truly hope that this book sets you solidly on the path to your own beautiful farm and a lifetime of green days and starry nights. We found it; you can too. — ALH

INTRODUCTION

The *perfect* organic farm for you is:

1. Located in a region with a climate and landscape you understand and enjoy.
2. Reasonably close to the farm markets and/or job opportunities you will need.
3. Reliably supplied with abundant, clean water year-round.
4. Blessed with enough farmable acreage, the right type of soil, good drainage, and terrain suited to the type of farming you plan to do.
5. Equipped with a soundly constructed house and outbuildings in good repair, or a good building site.
6. Available for sale with water and mineral rights included.
7. Situated in a neighborhood unspoiled by difficult neighbors, uncontrolled development, flooding, or environmental problems.
8. Appropriately regulated by government bodies that care about farmers.
9. Affordable for your budget.
10. Beautiful in your eyes.

The truth is that no property is perfect: there are always drawbacks. If you can't spot the problems — and the potential — of a farm for yourself, the next six chapters discuss what questions you'll need to ask (and why) to determine if a property could turn into your private purgatory, or has enough of the above attributes to become the beautiful small-scale organic farm of your dreams.

Finding good land, however, is only the first part of the equation. The second half — financing the purchase — is a major barrier for new farmers. There are no simple ways around this barrier, but you have options, including tapping into the expanding resources available to help new farmers. The final two chapters cover both traditional and creative methods for financing a land purchase.

In Recommended Reading on page 121 you'll find a list of publications that will assist you further in your search; you will certainly find more as your journey continues.

CHAPTER ONE

LOCATION

Why not begin searching for land now, even if you're not yet in a position to buy? Looking will clarify what you want for your farm and what is available.

The more you know about farming in general, your targeted locales, and how real estate ads compare to real estate reality, the better you'll be at spotting opportunities and avoiding problems. Drive rural neighborhoods, vacation where you hope someday to live, subscribe to the local paper, seek an internship or job on a farm, watch real estate ads, and go to small farm conferences and field days. Talk to everyone: real estate agents, Extension agents, farmers' market vendors, and the appliance repair guy. The Internet won't tell you which farms have poor wells or where the next big pig operation is getting built; the retired farmers hanging out at the local café might.

Commercial Farm Types

Aside from the subsistence farm, there are five basic types of commercial farms raising five types of products: vegetables and/or small fruits; tree fruits; field crops; dairy; meat or fiber animals (poultry, pigs, and grazing livestock).

There are many specialty crops as well, but in terms of what markets and land base they need, they will generally fit into one of these slots.

The type(s) you focus on will determine how much land and what sort of climate, soil, terrain, and rural neighborhood you will be seeking. You can, in fact, raise almost any category of farm product almost anywhere, but if you intend to make money, or at least not bleed yourself dry, you must have land appropriate to the enterprise and a reliable market where you can sell your products for more than it cost you to raise them.

Four Key Questions at the Start

As you look for land, define where and what kind of land you want by answering four questions:

1. **Do you want to generate most of your income on or off the farm?** Does access to off-farm employment or to good farm infrastructure and markets have priority?
2. **What would you like to produce?** If your intent is to sell farm products, you need land appropriate for your enterprise, and good markets.
3. **What are the other people involved in your farming venture looking for?** You may not care whether there are friends or a decent job close by, but your spouse or partner might. Children (whether present now or a future

possibility) might want to live close enough to town to avoid hours-long school bus rides.

4. **How will these needs and wants change in the long term?** Consider whether an area will also suit you when you are older — when adventure becomes less important and convenience more so.

What Farms Need

Different products require different soils, terrain, infrastructure, and markets. If you know already what you want to produce, use this chart to see what your farm will need.

PRODUCT	ACREAGE	TERRAIN	SOIL	WATER NEEDS
Vegetables, small fruits	0.5 acre (0.2 hectare) minimum	Level	Fertile, deep, well-drained	Abundant and reliable for irrigation system
Orchard fruits	5[+] acres (2[+] hectares)	Level to gently rolling	Moderately fertile, well-drained	Reliable, good quantity
Field crops	Moderate to major	Level to gently rolling	Reasonably fertile; well-drained	Varies by crop
Dairy	Moderate to major; must have pasture (to satisfy organic requirements)	Level to hilly if well managed	Adequate for pasture	Abundant and reliable
Poultry	Minimal	Level to rolling	Good for improving soil	Adequate
Pigs	Minimal	Level to rolling	Good for improving soil	Adequate
Sheep, goats, cattle	Moderate to major; must have pasture	Level to hilly (if well managed)	Good for improving soil	Adequate

PRODUCT	INFRASTRUCTURE	LABOR	MARKETS	START-UP COSTS
Vegetables, small fruits	Minimal	Intense during growing season; often extra help needed for harvest	Access to retail markets	Minimal
Orchard fruits	Minimal	Moderate; extra at harvest	Access to retail markets; processors for seconds	Moderate
Field crops	Moderate	Concentrated at planting and harvest	Access to brokers or livestock farms (feed)	Major: land, equipment, seed, soil amendments
Dairy	Major	Daily, year-round (unless seasonal)	Access to processors, or on-farm plant	Major: land, equipment, animals
Livestock	Moderate: processors, truckers	Low	Access to retail or middleman	Moderate: land, animals, fencing

DAIRY FARMS rely most on off-farm infrastructure, unless an on-farm plant processes the milk.

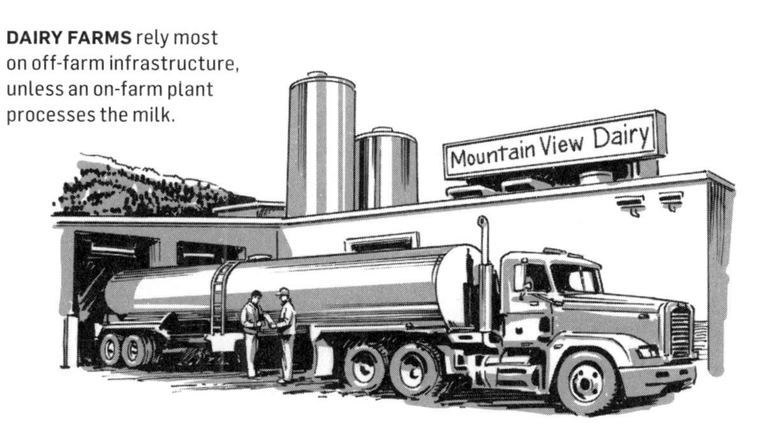

WHERE TO LOOK

IF YOU DON'T ALREADY KNOW where you want to farm, your answers to the above questions should help direct your search. The next step is to target a particular region.

Suitable Climate

You may already know the general area you're going to target in your search for land. You might already have a job, or want to stay close to familiar places and faces, or like a particular climate. But if you're open to living anywhere, the first thing to consider is the climate, for these three reasons:

Farming is an outdoor occupation and you'll be outside every day. Look for a region where you like most of the weather and can tolerate the rest.

Many fruits and field crops do better in some climates than in others. The plants and animals you plan to raise will be most productive where the climate is right for them. The USDA's National Agricultural Statistics Service site at www.nass.usda.gov is easy to navigate for state-by-state information on crops; state agriculture department websites and county Extension agents have useful information as well.

Climate determines the predominant farm types in a region, which indicates the availability of infrastructure. This is especially important if you're thinking about a field crop, large livestock, or dairy operation. Off-farm processing, trucking, storage, and distributors are most often needed by these types of farms.

Friends, Relatives, and Amenities

Access to decent schools, medical services, a good library, shopping, friends, and the relatives you like won't be critical to whether or not your farm is a success. But one of the big reasons you want your own farm is for the high quality of life, right? So it's important to consider how important those amenities are to your long-term happiness.

County by County: Markets and Jobs

Once you've targeted a state or two within your chosen climate region, narrow the search to the county level by going back to the first question on page 4: Will most of your cash income eventually come from the farm, or from off the farm? If the job comes first, then you'll need to have good access to the job. If the farm comes first, you'll need to have good access to markets appropriate for your product. Often the two coincide, but sometimes they don't.

IF THE JOB COMES FIRST

Statistically it is a rare farm, and an even rarer small-scale farm, where no one works off the farm. If all or most of your cash income will come from off the farm, then the best plan usually is to get the job first and then the farm. If you don't already live there, consider renting in the area while you search for land within commuting distance. If you can work at home, or have to go into the office only occasionally, then you can search a much wider area.

IF THE FARM COMES FIRST

If getting the right farm for your plans comes first, then identify the counties where there is both suitable land and good access to markets and infrastructure. The two charts on page 13 define what's needed by different types of farms.

How close you must be to those markets and infrastructure depends on how frequently you'll have to drive there, and how much time and gas you can afford. Be careful about overestimating your long-term tolerance for the driving; a four-hour trip to the weekly farmers' market may be tolerable for the first couple years, but after that it gets to be a bit much for most farmers.

A Few Rules of Thumb

Distance from customers can be critical. The rule of thumb is that every added mile and turn means fewer customers willing to drive regularly to your farm. A location close to town and on or near a major road is essential for on-farm sales.

Your market must be big enough. If you're planning on direct sales, you will need convenient access to a sizable pool of potential customers. Small towns don't usually generate a lot of sales. If you aren't close to a bigger town, consider looking for a middleman. Using a broker, distributor, or processor will cut your profit margins by quite a bit, but that will be balanced by your having more time to farm.

WINTER FARMERS' MARKETS are becoming more common, and a vigorous market in a nearby town greatly expands a small farm's ability to generate cash flow during the cold months.

Figure out what you can live with. There is always a drawback or a problem. Many problems can be tolerated or fixed, but if there are three or four big items lacking, that is probably a good reason to look somewhere else for a farm.

Drawbacks vs. Deal-breakers

Once you've learned to identify the drawbacks of a property, the next step is to separate deal-breakers from negotiating points. Deal-breakers are unacceptable problems with a property: either you can't fix them or you can't afford to fix them, and you refuse to live with them. Negotiating points are those things that you can live with or you can fix. The prospect of dealing with these drawbacks, however, will influence how much you are willing to pay and how soon you will have the time and money to turn most of your attention to where it belongs — farming. Which items appear in which category will differ from buyer to buyer: one new farmer's deal-breaker may be another's opportunity to apply some elbow grease and creativity to reveal a hidden gem.

HOW TO LOOK

It's essential to investigate widely and creatively as you search for rural properties for sale. Don't rely exclusively on the Internet: many sellers are older people who don't use computers much and prefer more traditional channels.

Common Marketing Methods

FARM CATEGORY	VEGETABLES/ SMALL FRUITS	TREE FRUITS/ NUTS	FIELD CROPS	DAIRY	MEAT
Farmers' Markets	x	x	—	—	x
CSA	x	—	x (small grains)	—	—
Direct to Institutions	x	—	—	—	—
On-Farm Sales	x	x	— (web-based)	x	x
Sale through Middleman	x	x	x	x	x
Delivery to Individuals	x			—	x

x = it's very common; — = it's being done

Desirable Nearby Infrastructure

FARM CATEGORY	VEGETABLES/ SMALL FRUITS	TREE FRUITS/ NUTS	FIELD CROPS	DAIRY	MEAT
Breeding Stock				x	x
Veterinarian				x	x
Equipment Dealer	x	x	x	x	x
Farm Store	x	x	x	x	x
Feed Supplier				x	x
Trucking				x	x
Processing				x	x

How Property Is Put Up for Sale

Property is advertised and sold through a variety of channels:

"For Sale by Owner" (FSBO) ads are found on the Internet, in local papers, and often those of nearby larger towns and cities. The owner may stick a sign in the yard as well, or do nothing at all to advertise, simply passing the word among friends and relatives. If you see a promising farm with no "For Sale" sign, you might politely ask the owners if they are considering selling. Property frequently changes hands in the country by word-of-mouth, never appearing on the market.

Real estate agents and brokers are often your best bet for finding the right farm. Though legally they work for the seller of the property, and their fee comes out of the sale price, a good real estate agent will work hard for you as well. Check listings at different real estate agencies and especially the locally owned ones that don't have national or regional affiliations. The independent agents may know the area better.

Land auctions can be conducted by a county when property taxes are in arrears, by the Internal Revenue Service for other delinquencies or crimes, by U.S. Bankruptcy Court, and by banks and credit unions foreclosing for failure to make mortgage payments. Owners, estates, and business entities such as timber companies may auction land as well. *Be sure to research the protocols and legal requirements* for each type of auction before participating, and never, never bid on land that you haven't seen! Consult expert advice on any legal concerns. Some useful books covering this topic are listed in Recommended Reading, page 121.

Developers of rural properties may offer "large lots" of 20 acres and more. Be sure to do your homework when looking at this type of property; buyers are often restricted in what they can do with their land. Other pitfalls include poorly built roads (for more on this, see chapter 6), inadequate surveys that leave property lines in doubt, and the chance of the developer going bankrupt before promised improvements are finished.

5 Places to Look for Property Listings

Once you've defined where you are looking for a farm, it's time to start looking at ads and making contacts. You can find property listings by:

1. **Checking organic and sustainable agriculture websites** that have classified ads — these sometimes include farms for sale. Start with the website National Farm Transition Network at www.farmtransition.org, a clearinghouse for state sites that link retiring farmers to new farmer's.
2. **Checking the local classifieds,** such as Craigslist and other regional electronic want-ad websites, and local print media including newspapers and advertising "trading post"–type circulars. If the region or state has farm-specific newspapers or magazines, get a subscription. Find these by Googling "agricultural newspapers" with the name of the state.
3. **Contacting local real estate agents** and telling them what you're looking for. The American Society of Farm

Managers and Rural Appraisers at www.asfmra.org can connect you with real estate agents who are also appraisers, a nice combination.

4. **Checking county government and local bank websites** for listings of foreclosed properties. The U.S. Department of Agriculture maintains a national "clearing house" website at www.resales.usda.gov for government land auctions and properties for sale, which includes farms and rural acreage. Agriseek at www.agriseek.com lists government and bank foreclosures as well as land available for rent or lease. Government Auctions at www.governmentauctions.org has state-by-state foreclosure listings that include land. Auction.com at www.auction.com lists government and bank foreclosures, mostly residential but some acreages.

5. **Watching Internet rural real estate listings.** Start with:

LandAndFarm.com
www.landandfarm.com

LandsofAmerica.com
www.landsofamerica.com

RuralProperty.net
www.ruralproperty.net

United Country
Real Estate
www.unitedcountry.com

RanchandCountry.com
www.ranchandcountry.com

Rural Property Finder
www.ruralpropertyfinder.com

U. S. Land and Home
www.uslandandhome.com

HOW TO READ A FARM REAL ESTATE AD

You may encounter some unfamiliar terms when you first begin reading farm real estate ads:

- **Tillable, arable.** Describes land (usually stated in acres) that can be worked with machinery and used for growing field or garden crops.
- **Open.** Describes the number of acres that are not wooded but not suitable for tillage; usually this means pasture, and not always good pasture.
- **Wooded.** Describes the number of acres with trees. This does not give you any indication of species, age, value, or number of trees.
- **Other.** This means wetland, rock, brush, or some other type of acreage that is unusable for farming.

Real estate ads rarely match real estate reality. Never make an offer on a property without seeing it a few times and doing your homework on the soil, water supply, soundness of the buildings, neighborhood, and local regulations!

CHAPTER TWO

WATER

The very first thing to look at when visiting land for sale is the water supply. If there isn't enough, or it's contaminated, or it dries up every summer, then you won't be able to farm.

Water problems are common in rural areas and have numerous causes. If there is a problem with the water supply on a property you're really interested in buying, you must determine if you can live with the problem or fix it, or if it's time to walk away.

Many state health departments and/or departments of natural resources have excellent general information on private water systems (a few are mentioned below), as well as the most common or intractable problems in the state, such as atrazine contamination in some Wisconsin wells. Be sure to check the website for your state, or stop by the local office.

If you are considering buying undeveloped land, the sale should be contingent on whether the land has an adequate supply of good-quality water. This may involve drilling a test bore to see how deep it is to water, which may cost several hundred dollars or more. Then you can get an estimate from the driller for how much it will cost to install a well. Do not accept an owner's or real estate agent's verbal guarantee that there is water — make the contingency part of the written sales contract, so if no water is found, you won't be left paying for land you can't farm.

Assessing the Water Supply

Hand pump for water well

1. Determine the type and age of the system. If a well, obtain the well driller's report from the County. If another type of system, ask the owner to find out how old it is.

2. Have the water tested for quality. Labs can be found through the state health department.

3. Assess quantity by getting a seasonal history and by running several taps for several minutes to see if flow rate is maintained.

WHAT'S THE SOURCE?

THE FIRST STEP WHEN ASSESSING a water supply is seeing where the water comes from, and if that source meets state water quality standards. Rural water can be supplied by a well, spring, surface waters, rainwater and storage cistern, or a public system. Find out how old the system is since this will give you a notion of how soon it might need repair, renovation, or replacement.

Wells

The most common source of water in rural areas, and the most likely to be clean and reliable, is a well. There are three types: dug, driven, and drilled. According to U.S. Environmental Protection Agency standards (state and local standards may vary), all should be located:

- At least 50 feet (15 m) from septic tanks, leach fields, livestock yards, and silos
- At least 100 feet (30 m) from any petroleum tanks
- At least 250 feet (76 m) from manure storage

A well too close to any of these contamination sources is suspect. The well should also be *uphill* from the barnyard and the septic system.

All types of wells should have a watertight casing rising at least a foot (0.3 m) above the ground, and be securely capped. There should be no cracks or wiggle in the casing and the ground around it should slope away at least slightly. Dug and driven wells have external surface pumps; this should be inside the house or in a secure well house.

FIXING WELL PROBLEMS

Wells don't last forever: the intake screen at the bottom of the pipe may clog or rust up, the pipe may rust out, the pump may go out or leak lubricant into the well, or the groundwater source may run dry. This could happen over a couple decades,

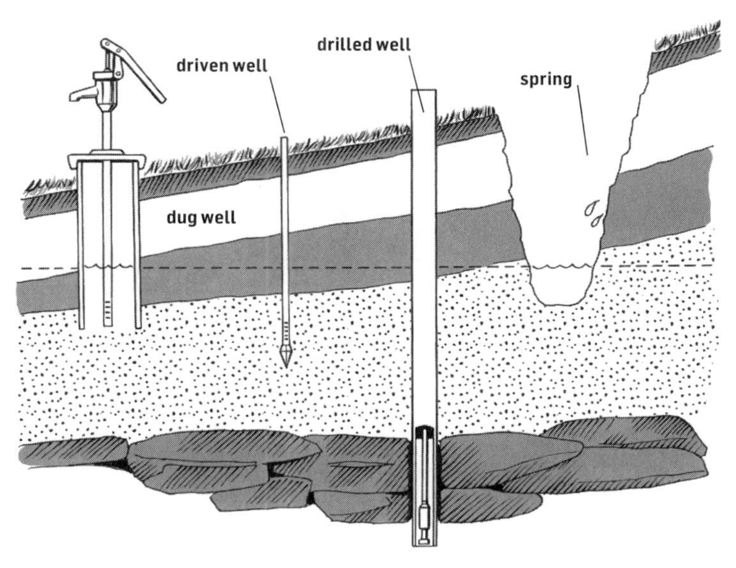

DUG WELLS tend to be wide and shallow (10 to 30 feet), and may be more likely to go dry during a drought. **Driven wells** average 30 to 50 feet (10–15 m), and the pipe 1 to 3 inches in diameter. **Drilled wells** may go as deep as several hundred feet, and typically have a submersible pump at the bottom.

or you might get lucky like us and the original well will last more than 50 years. The fix may be simple (such as replacing the pump), complex and costly (such as putting in a new well), or downright unsolvable (the water might be gone). If there is

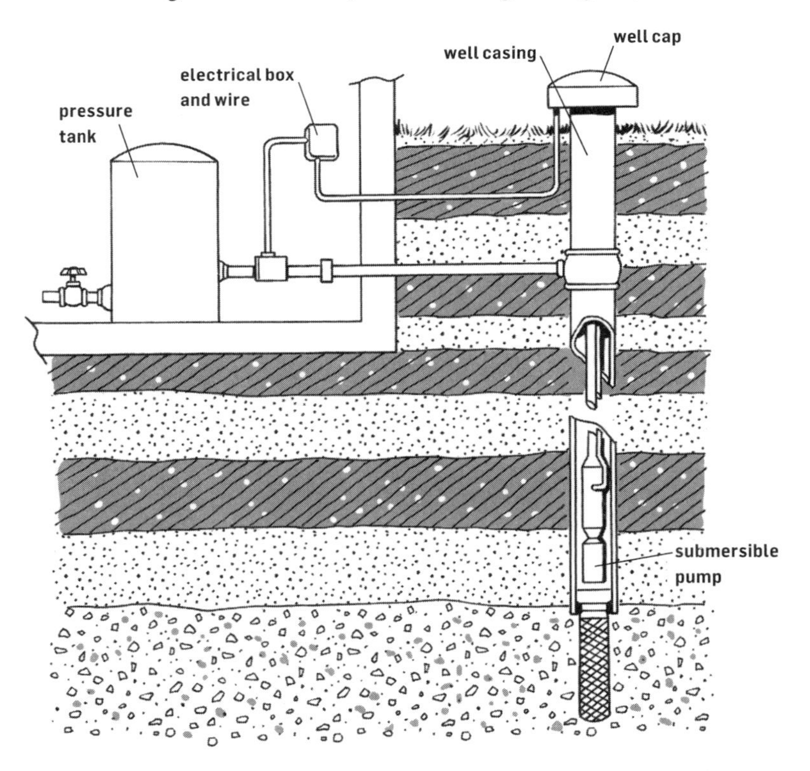

DRILLED WELLS with submersible pumps are the most common type of well in many areas. A pressure tank inside the basement maintains constant water pressure.

any sign of a problem, it is important to diagnose it and, if the solution will be expensive, include the cost in your negotiations with the property owner.

If, for example, the current water flow is quite a bit slower than what was measured when the well was first put in (the original flow rate should be part of the well driller's report — see page 24), that's a sign that something is not right. When our water flow slowed to less than 5 gallons (19 L) per minute, we ultimately had to put in a new well, 17 years after farm purchase. When we called a well driller, he diagnosed a clogged intake screen at the bottom of the well. This meant that the water supply was fine, fortunately, and we could put the new well in close to where the old one had been.

COST

Well construction costs usually run by the foot. Your cost will depend on market rates charged in your location, the type of well, how deep the well is, and whether the pipe has to go through rock or soil. To give you some idea, our new 115-foot-deep (35 m) well, with 6-inch (15 cm) pipe and submersible pump, and including capping, sealing, disinfection, water testing, and sealing of the old well, cost us $5,008 in 2009. Our driller's rate was $32 per foot plus all the other items. We now get a steady 20 gallons (76 L) per minute.

POSSIBLE RED FLAG

Old wells are often abandoned after the landowners put in a new, deeper well; if you find the pipe or pit that marks an old well, make sure it has been sealed according to local regulations. This may have to be done by a licensed well driller, but is not usually too expensive — though it's always wise to get a quote on the price. Abandoned wells are not only a hazard to people and animals, but they also provide a direct conduit for contaminants to reach the groundwater without filtering through the soil.

THE WELL DRILLER'S REPORT

Wells have been regulated at the state and local level in most areas for many decades. In most states, you will need a professional driller and a permit from the local authorities to establish a well. The well driller is required to file a report with the county.

If you are seriously interested in a property, get a copy of the well driller's report from the county seat. It may be the department of health, natural resources, or something else; just ask. The well driller's report will tell you:

- The year the well was put in
- The pipe diameter
- The depth to water and the depth of the well (the bottom of the well should extend well into the water table)
- That the well was disinfected and sealed upon completion
- That the water was laboratory tested and certified as safe

Springs

Springs are water sources in some areas. The same general considerations apply for springs as for wells:

- Is the spring protected from contamination?
- Is there enough water?
- Does it run reliably throughout the year?

North Carolina Cooperative Extension recommends that spring water be intercepted underground and piped to a sanitary spring box with a tight cover to provide maximum protection against contamination.

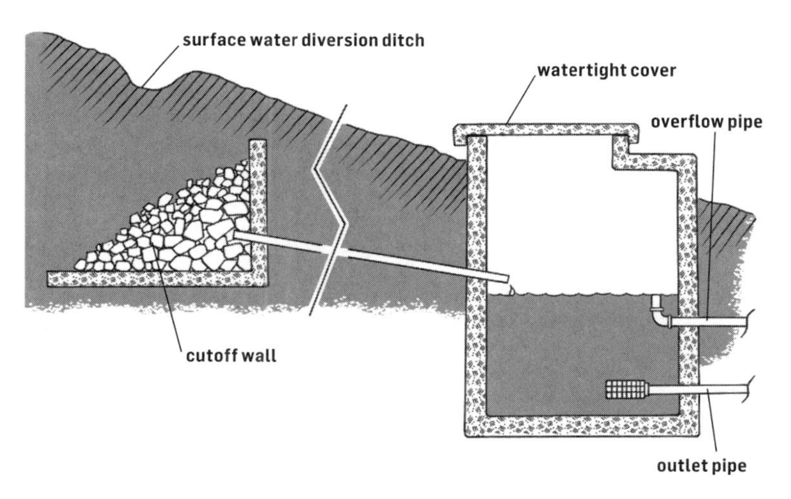

SPRINGS THAT EMERGE from hillsides are the easiest type to develop. The water is intercepted by an underground cutoff wall and piped to a sealed tank in order to prevent contamination from litter or surface runoff.

Most important is that there is nothing uphill from the spring — such as big livestock farms, mines, or logging operations — that could contaminate the groundwater flowing into the spring. To get a notion of what is uphill, you could take a drive around the neighborhood, or even better, use Google Earth (download free from www.googleearth.com) to really see what's behind those trees or hills and down those dead-end roads.

Surface Waters

If the farm relies on water from either a pond or stream, there should be a system in place to draw the water, sanitize it, and store it for use. The University of Missouri Extension website's http://extension.missouri.edu/publications page recommends that the water be piped to a compartment where it is filtered through sand to a gravel bed to remove foreign matter, then piped to another storage compartment where it is treated with chlorine to kill bacteria.

If the source is moving water — a stream or river — there will most likely be restrictions on how much water you can draw. In the East, this may not be an issue except in a severe drought; in the West, it will certainly be regulated.

Cisterns

Where ground and surface waters are inadequate or unavailable, a cistern can store rainwater (or trucked-in water if on-farm sources fail or become so contaminated as to be unusable).

The cistern should be sealed tightly to prevent contamination, and sized for the farm's needs.

You can calculate the potential gallons collected in a cistern by multiplying the roof area by the average inches of rain per year; for example, 7 inches (18 cm) of rain on 1,500 square feet (140 sq m) of roof yields about 6,300 gallons (23,848 L), according to the Ohio Department of Health. If the property you're considering relies on rainwater, keep in mind that drought occurs in all areas of the country; you'll want a generous-sized cistern and perhaps the ability to haul water from a nearby town if things get really dry.

Ownership

Water rights may be separate from the title to a property; this is common in the western United States. Before buying property, it is essential that you ensure two things in writing: you will be able to purchase the rights to enough good water for your farm; and if not, the sale will be voided.

In some areas (such as Colorado) you may not even have the legal right to collect the rainwater that falls on your property if your farm does not meet regulatory criteria. In fact, you may not even be able to reuse your gray water in Colorado, depending on the terms of your well permit.

Since the laws and entities governing water rights can be complex and vary from state to state, get *expert* advice on this topic in any area (particularly west of the Mississippi River) where water rights do not run with the land.

Public Water Supplies

A public water supply (piped, or sometimes trucked) is available in some rural areas. If land you're considering purchasing has access to a public water supply, inquire about the charges for use. In this case, the public authorities will take care of quantity, reliability, and quality. If no public water is available, however, it's highly likely that it never will be, so don't count on that when making your farm plans. Our township regularly gets inquiries about how soon we'll be getting municipal water and sewer, and the answer is "never."

AMOUNT AND RELIABILITY

ESTIMATE WHETHER THERE IS ADEQUATE WATER on a property by adding the daily requirements of the humans, livestock, and plants that you plan to grow, then comparing the total to what is available (see below, Is There Enough Water?). Keep in mind that many water sources dwindle during dry spells. A long and severe drought will affect almost any water source, but a normal dry spell in late summer shouldn't leave you thirsty. You'll want to check the water supply in a dry period if possible, and err on the generous side when anticipating water requirements. If the water supply greatly exceeds the need, count yourself lucky: you can look forward to a reliable shower after being outside all day.

Is There Enough Water?

If the well delivers more than 1 gallon (4 L) per minute, or 1,440 gallons (5,450 L) per day, with care your water should be adequate for your needs, but there won't be much wiggle room if you get into a drought or have a lot of company over the weekend. If the well delivers 10 gallons (38 L) per minute, or 14,400 gallons (54,510 L) per day (assuming the water supply will tolerate continuous pumping), then you can think about adding a goat herd and apple orchard to your operation! Here's how to calculate whether the water supply is adequate to meet your needs.

DAILY WATER USE

Household. Average U.S. daily household water consumption is about 70 gallons (265 L) per person.

Livestock. Livestock consume more water in dry seasons — including winter in northern areas — and when milking. Average daily water requirements:

- Beef cattle: 5–15 gallons (19–57 L) each
- Dairy cows: 35–45 gallons (132–170 L) each. Dairy operations require additional water to clean milking equipment
- Pigs: 1–2 gallons (4–8 L) each
- Sheep and goats: about 2 gallons (8 L) each
- Chickens: 1–2 gallons per 10 birds (4–8 L)

Vegetables. Average daily water requirements: About 3,900 gallons (14,763 L) per acre (0.4 hectare) per day, assuming an average weekly requirement of 1 inch (2.5 cm) of rain.

ADD IT UP

For example, if you have:

Two adults using 70 gallons (265 L) each	=	140 gallons (530 L)
{+} 10 chickens	=	2 gallons (8 L)
{+} ¼-acre (0.1 hectare) garden	=	975 gallons (3,691 L)

You will need: 1,117 gallons (4,228 L) of water per day.

Water Flow and Pressure

Measuring the flow rate of a water system is not an exact science, and with all but deep wells the rate may vary quite a bit through the year. The type of farm may be a good clue to how much water is available: dairies and vegetable operations, for example, use a lot of water. Talk to the owner, the neighbors, or the Extension agent and review the well driller's report to get an idea of the amount and reliability of the supply.

If there's really no information, you can get a rough idea with a 5-gallon (19 L) bucket. Since it's common to have a pressure tank in the basement to equalize water pressure in the house, either find an outside hydrant that isn't run through the pressure tank, or let one — or several — inside faucets run for a while to deplete the tank. Then time how long it takes to fill a 5-gallon bucket, and calculate how many gallons a minute that would equal.

Another thing to check is whether the water pressure is as good upstairs as it is in the basement; if it's not, the problem is usually not the source, but old pipes clogged with mineral deposits.

TESTING QUALITY

IF YOU ARE SERIOUS ABOUT A PROPERTY, take a water sample and have the water tested by a certified lab for *E. coli* (*Escherichia coli*) bacteria, nitrates, hardness, alkalinity, pH, conductivity, and chloride. Make sure you draw the sample from a faucet that is not hooked up to a water softener or any other sort of filtering or treatment system.

The state department of health or natural resources website will have information on how to find a lab. The EPA site at http://water.epa.gov also has excellent information on water quality issues as well as links to state sites. If the state website also recommends specific other tests (such as for atrazine in Wisconsin), order these as well.

If the test comes back clean, there are probably no serious problems with the water quality. If tests indicate *E. coli* or nitrate levels are above acceptable limits, you should do some additional testing, since those results may indicate the presence of additional contaminants (see chart on page 32). You'll also need to determine if the source of the problem is on the property or somewhere upstream; this will involve, for starters, a close (and maybe professional) inspection of the septic system. If the chloride level is high, the source may be road salts or mining operations in the area.

Contaminants Travel in Packs

IF THIS CONTAMINANT IS PRESENT:	THEN ALSO TEST FOR:	POSSIBLE SOURCES:
E. coli bacteria (not normally disease causing)	Cryptosporidium, Giardia lamblia, other disease-causing bacteria	Septic system, manure
Nitrate	Pesticides, and arsenic if there are intensive poultry or orchard operations in the area	Septic field, manure, commercial chicken feed, synthetic ag chemicals

If tests for water hardness, alkalinity, pH, and conductivity come back with poor results, the cause will most often be either the natural state of the water or the plumbing. These problems are common and, though they may be unpleasant, won't affect your health and so are not normally deal-breakers. The solution is usually a treatment system (a water softener is the most common in my area). Tests that reveal lead in the water, however, *will* affect your health, and you will need to replace the old plumbing or install a treatment system to make the water safe.

When you take the sample for testing, you should also put some water in a clear glass and hold it up to the light to see if it's clear, then smell and taste it. Look at the pipes to see if they are corroded, or old enough to have lead parts or solder. That will give you an initial clue to some of the more common problems. Poor results from these inspections or those previously discussed above indicate a need for additional water testing. For further guidance see the page on drinking water contaminants at the EPA website (see page 31).

You can also test for polychlorinated biphenols (PCBs) if the water source is a river or stream, since PCBs linger in stream sediments. If there is a chemical waste dump on the property (look for old containers), an underground fuel storage tank (common on many larger old farms), or an uncontained chemical mixing area (ask where they filled up the sprayer tanks and changed oil and hydraulic fluid), have the water tested for petroleum products and other chemicals.

Contaminated Water and Organic Certification

According to Harriet Behar, organic specialist at the Midwest Organic and Sustainable Education Service (www.mosesorganic.org), the National Organic Rule is mute on the quality of water used for plants or livestock on a certified organic farm (although this could change in the future). The rule does state that only potable drinking water can come in contact with organic food, but this is when washing the produce after harvest, not out in the field.

For the farmers' own safety and that of their customers and livestock, Behar says, they should not use — for *any* purpose — any water that could cause damage to human health due to contamination with bacteria, fertilizers, manure, or herbicides. A further danger of using contaminated water is that it could expose the farmer to legal action if a consumer became ill as a result of such use.

Various states regulate the water supply of dairy animals; check what regulations are in force in your area by calling your state department of agriculture. The absence of regulations, though, doesn't make it okay to use bad water on a good animal, Behar emphasizes. If the animal gets sick, the farmer loses money either trying to get the animal healthy again, or through the loss of the animal when it dies.

Smart choices about the quality of your water source are more about common sense than government regulations, Behar says.

Dealing with Water Contamination

If the water is contaminated, the first step is to determine the source of the problem: the well, the plumbing, an on-farm source, or an off-farm source. As a general rule, you should fix problems whose source is on-site; off-site problems generally require installation of a treatment system capable of dealing with the specific contaminant(s). In some cases, there is no good remedy.

If the water is contaminated with bacteria, you can start by shock-chlorinating the well. Follow the directions on the EPA or the American Ground Water Trust website (www.agwt.org), or hire a professional. If this doesn't provide a permanent fix, the problem probably comes from beyond the well.

- If the source is on the property, you may have to reconstruct all or part of the septic system, or drill a new well at a location that won't be impacted by runoff or leachate from the barnyard, septic system, or former petroleum and ag-chemical storage and mixing areas.

- If the problem comes from off-site, you may have to treat the water at home or bring it in from an outside source for the foreseeable future. Though filters are not effective against bacteria, boiling will kill most bacteria. Home treatment systems utilizing ultraviolet light, iodine, ozone, or chlorine are also effective.

If the water is contaminated with nitrate, the water is unfit to use for drinking or for livestock, and can cause serious health problems in infants. Home treatment systems for this contaminant may utilize anion exchange, reverse osmosis, or distillation.

If the water is contaminated with lead, the problem probably comes not from indoor plumbing, but from the old buried pipe that runs from the well or other water source to the house. Replacing the pipe is messy and may cost several hundred dollars or more, but it should cure the problem permanently.

If the water is so hard that you can't get your soap to suds in the shower, or your white clothes begin looking gray, then consider renting or buying a water softener. If there are high iron levels in the water as well, you can add "iron out" to the softener salt.

For more information on water treatment options, visit the American Ground Water Trust website for an excellent introduction to home treatment systems for bacteria, nitrates, and other contaminants.

CHAPTER THREE

GOOD LAND

To a farmer, the word "land" means the sum of all its parts, which together determine what kind of farming will suit that particular place. These parts — or characteristics — are: soil type, potential fertility, and current condition; how well it is drained; which way the land faces and how steeply it slopes; and whether or not the fields are easy and efficient to work.

Matching Farmland to Farm Goals

The first question to ask as you walk a property is if it suits the type of farming that *you* plan to do:

Vegetables and small fruits need minimal acreage but level land with deep, fertile soil and good drainage.

Orchard fruits and nut trees need more room than vegetables but are not quite as demanding of the soil. Moderately sloped or rolling land is preferable, and never at the bottom of a valley or other potential "frost pocket." The ideal is moderately to very fertile, deep soil with good drainage.

Field crops require level to moderately rolling land, with moderately to very fertile soil and good drainage. For commercial production, field crops require considerably more acreage than do orchards.

Dairy animals require level to moderately rolling land for hay and grain production, and can use rough land for pasture as long as it's at least moderately fertile and capable of growing good forages — though level, fertile pasture works well, too. The land must be well drained with enough acres to provide pasture

WHEN LOOKING AT LAND TO BUY, walk all the boundaries and through all the fields (with a shovel or soil auger and bucket if you're going to pull soil samples), to get a detailed idea of slope, aspect, drainage, field layouts, and soil type and fertility.

and hay. The acres-per-head ratio varies widely by region; for specifics, ask area farmers or the Extension agent.

Meat livestock can utilize rough and poor land except for the wettest areas or steepest slopes, as long as some palatable forage grows on it. If you plan to grow hay for feeding in winter or dry seasons, you'll need some level to moderately rolling land with decent fertility. Ask local farmers or the county Extension agent for a ballpark figure of how many acres you'll need per head of livestock.

In modern times, agricultural academics have refined our collective knowledge of the different characteristics of land into technical descriptions and calculations. This information is now easily available in the form of soil-type maps and descriptions and land-capability class tables. For those qualities that are not easily categorized — such as the aspect or the field layout, or whether a degraded soil is worth trying to restore on your budget — you will have to rely on the information you can find to decide for yourself.

SOIL IN GOOD HEART

Good soil is the foundation of a farm, and the organic farmer is explicitly charged by the Organic Rule to constantly preserve and build soil fertility. To do this requires a basic knowledge of soil, and when better to start learning than when you're looking at farmland? Knowledge is especially important if you are trying to decide whether degraded soil on a property can be restored.

Whether or not a soil is good for farming depends on its type, capability class, and the gap between its current fertility and its potential. Degraded soils are common, and some can't be restored — or could be restored but not in any way that makes economic sense (see pages 45–46).

On the other hand, as a small-scale, creative, organic farmer, you may be able to restore some abused soils. Utilizing organic and sustainable systems and practices — and there are many — you can turn once-good, now less-than-fertile soil into healthy, productive soil, or "soil in good heart."

ROCKS AND EQUIPMENT DON'T MIX

A lot of rock on and in the soil breaks equipment and tools. With a strong back you can manually pick rocks on a small acreage with a moderate rock load. Larger acreages can sometimes be cleared with hired equipment, if the land isn't too steep.

Soil Type

Soil comes in three broad classes, depending on the ratio of coarse, fine, and medium-sized mineral particles in its makeup.

Sandy soils are coarse, light, and easy to work. These soils warm up quickly in the spring and are especially good for root crops and deep-rooted plants such as alfalfa. Sandy soils do not hold water or nutrients well and so require more constant attention to maintain fertility.

Loamy soils are medium-bodied, and the best all-around soil for almost any type of farm production. When well-tended, these soils consistently produce high yields.

Clay soils are highly fertile, but heavy and sticky to work. They warm up slowly in the spring but hold water longer into a drought than do other soils. Clays are poor soils for root crops, but good for crops that are most susceptible to drought.

Each of the broad soil categories is subdivided into precisely defined types. To get a map of the types on a specific property, either request one from the county Extension agent, or go to the USDA's Natural Resources Conservation Service's National Cooperative Soil Survey at http://websoilsurvey.nrcs.usda.gov to create and print out a customized map.

DEGRADED SOILS with good potential for restoration can be a rewarding project for the organic farmer with some time and spare room in the budget. Rotationally grazed livestock add fertility to the soil with their manure, while the grasses they graze build soil structure and organic matter with their roots. Plowing under green manure crops adds nutrients and organic matter, while correct tillage — in this case, across the slope rather than up and down — prevents erosion.

Look up the description for each soil type (it might be easier to find at the Extension office than on the website), and pick out the following important points: depth of both topsoil and subsoil, drainage, degree of slope, and suitable crops for the soil. The description usually ends with the soil's capability class, which adds up a soil's basic type, depth, slope, and other factors to a single estimate of its degree of farmability. With this information in hand, you will have a good idea of a property's potential, but you will still need to walk the land and have a soil test done to determine its current fertility.

LAND CLASSES AND DEFINITIONS

Farmland is classified into capability classes by the Natural Resources Conservation Service according to its ability to sustain farm production. Generally, classes I to IV are considered farmable, though the higher the class number the more "good management and conservation treatment" will be required to maintain soil productivity.

- **Class I** soils have slight limitations that restrict their use.
- **Class II** soils have moderate limitations that reduce the choice of plants or require moderate conservation practices.
- **Class III** soils have severe limitations that reduce the choice of plants or require special conservation practices, or both.
- **Class IV** soils have very severe limitations that restrict the choice of plants or require very careful management, or both.

- **Class V** soils have little or no hazard of erosion but have other limitations, impractical to remove, that restrict their use mainly to pasture, range, forestland, or wildlife food and cover.
- **Class VI** soils have severe limitations that make them generally unsuited to cultivation and that restrict their use mainly to pasture, range, forestland, or wildlife food and cover.
- **Class VII** soils have very severe limitations that make them unsuited to cultivation and that restrict their use mainly to grazing, forestland, or wildlife.
- **Class VIII** soils and miscellaneous areas have limitations that preclude their use for commercial plant production and that restrict their use to recreation, wildlife, or water supply or for aesthetic purposes.

Subclasses. In addition, each capability class can be subdivided into subclasses. Land capability subclasses are the key to the most immediate problems with a soil:

- **Subclass "e"** soils are prone to erosion and may already have erosion damage.
- **Subclass "w"** soils suffer from excess water. The problem could be poor soil drainage, a high water table, or the fact that it's in a floodplain.
- **Subclass "s"** soils have problems in the rooting zone, which may be anything from shallowness of the topsoil, to an abundance of stones, low moisture-holding capacity (droughty), low fertility that is difficult to correct, or high salinity or sodium.
- **Subclass "c"** soils are limited by climate (either extreme temperatures or lack of moisture).

Fertility: Time to Walk the Land

Assessing the current state of the soil begins with walking the land.

First look at whatever is growing on the land now: crops, brush, grass, trees. The lushness of the growth is a good indicator of soil fertility. The vigor and variety of native plant species are also good indicators of soil type and quality.

Look for signs of erosion and overgrazing such as gullies, poor or no topsoil, lots of bare or nearly bare dirt in pastures, muddy creeks and ponds, and thistles standing tall like specimen plants while the grass is grazed down to the dirt. Dead and dying trees in pastures usually indicate soil compaction.

Testing Soil

At this point, if you're seriously interested in the land, go for another walk, taking a bucket and a shovel, trowel, or soil auger borrowed from the Extension office. Stick your implement in the ground here and there, and examine the soil: its color, smell, texture, and the depth of the topsoil.

Good dirt, with good levels of organic matter and nutrients, should be a deep, rich color and smell almost sweet, not sour, bitter, musty, or moldy. Bugs, worms, organic debris, and big. deep plant roots are all good signs of a living soil. A handful of moist soil should squeeze into a ball, then break at a touch into smaller clumps, not stay in a ball or crumble to dust.

Next take a soil sample. This typically involves mixing clean topsoil from several different points around a field, bagging it, and sending it off to the soil lab for analysis. Use a lab in the state or at least the region, one that is familiar with the local soil types. Labs can be found through your county Extension agent, or with an online search. For a reasonable fee, your county Extension office may take care of getting the sample to the lab for you and can help interpret the results. Call ahead of time; most labs will mail you bags for the samples with their instructions, or you can often pick up sample bags at the Extension office.

Normally soils are analyzed for pH, organic matter percentage, nitrogen, phosphorus, potassium, calcium, and various micronutrients. The analysis includes recommendations for the soil, according to the crops you intend to plant. Do not settle for less than this panoply of tests; they will indicate how much you may have to invest to realize the soil's full potential.

If the test results show serious deficiencies, consult a soil specialist, preferably one who is knowledgeable about organic agriculture, to discuss the methods and costs of making the soil productive.

Restoring the Soil

Much can be done to put the heart back into soil by tilling in green manures, compost, free organic matter (leaves from nearby towns, for example), and animal manures. Some inputs may have to be purchased to correct macro- or micronutrient deficiencies or pH; prices will vary by year and region. In a very small area you can till in amendments with a shovel; for more acres you'll need some equipment. Smaller areas are cheaper and easier to restore than larger acreages.

CAN THIS SOIL BE SAVED?

Generally, the following soil situations can be remedied:

- **Soils in land capability classes I through IV and subclass "e"** (see pages 41–42) can potentially be restored at a reasonable cost, if the erosion is not so severe that the topsoil is completely gone, or that deep gullies are making fields unusable.
- **Soils in subclass "w" need drainage to be usable.** This can work as long as there is somewhere lower for the water to go, and if ditches or subsurface drainage can be installed, an expensive proposition for any sizable acreage.

If you see the following, it may never be possible to make the land suitable for farming:

- **Soil too shallow** (6 inches [15 cm] or less) to support productive crops due to rock, or natural hardpan (subclass "s"), or because of a high water table (subclass "w"). ***How to identify:*** This is characterized by sparse, stunted vegetation in the case of rock and hardpan. Soil with a high water table planted with normal crops will show the same symptom; if wetland plants are growing there instead they may appear green and lush. Dig around with a shovel to see if this is indeed the problem.
- **Salinized land**, where overirrigation has left so much salt in the root zone that many plants won't grow, and those that do, struggle. ***How to identify:*** The certain diagnosis of this condition is a soil test.
- **Soil so naturally acid or alkaline** that it will not grow farm crops. ***How to identify:*** Test the soil. Neutral soil has a

pH of 7. Soils with a pH ranging from about 6 to 7.5 can usually be corrected; beyond that, long-term costs should be carefully considered. Or plan to grow specialty crops such as blueberries, which require a more acidic soil.

- **Wet land** with no potential for drainage. ***How to identify:*** Look for the lush, coarse vegetation characteristic of wetlands; the ground may also be soggy underfoot.
- **Land with poisoned water** from agricultural chemicals or other pollutants. ***How to identify:*** This is discovered by testing the water.
- **Severely degraded sites,** such as reclaimed strip mines, badly eroded lands, shale banks, and arid overgrazed areas. ***How to identify:*** Characterized by little or no topsoil and poor-quality vegetation, these sites may also have a large amount of undesirable and/or invasive plant species.

Hardpan

Hardpan or claypan — an impermeable soil layer below the topsoil — can prevent root penetration and/or hold enough water in a soil to stunt growth. These layers can form naturally or be caused by years of heavy equipment use, or always tilling to the same depth. Some hardpan can be corrected by tilling with a "subsoiler," although this can be fairly expensive to hire.

SLOPE AND ASPECT

THE STEEPNESS OF SLOPES and which direction they face — their aspect — are essential factors in farming.

The steeper the slope, the harder it is to use equipment, and the more measures you have to take to prevent erosion, that killer of farmland. Farmers produce crops on fairly steep slopes all around the world, but more sustainable farms utilize contour cropping, terraces, and other techniques to keep the land from washing away. (Land capability classes include the degree of slope; see above.) The steeper the slope, the more pronounced the effect of aspect will be.

Aspect matters for a variety of reasons, as discussed below. There are, in fact, no "bad" aspects: all have advantages and disadvantages. The important thing is how the direction the land faces will suit the type of farming you intend to do.

Facing south. In the spring, south-facing land will thaw, dry, and warm enough for planting or pasture growth anywhere from a few days to a couple of weeks earlier than a north slope will. A sheltered south slope may thus make it possible to grow longer-season vegetables or begin grazing earlier in the season.

Facing north. A northern slope will be cooler and moister than a southern slope. In a dry spell, plants on a north-facing slope (if the soil has been tended properly) will continue to grow well for some time after plants on the south side have given up and gone dormant.

If you want to grow fruit tree varieties that bloom before the average last frost date in the area, planting on a north slope may delay blossom time long enough to avoid a freeze killing the blossoms and ending any hope of a good harvest that year. But you may have to plant shorter-season varieties.

Facing east. East-facing slopes warm up more quickly in the morning and are cooler in the afternoon — good for helping honey bees get an early start and grazing animals get out of the hot sun.

Facing west. Western slopes get hotter in the afternoon but don't receive quite the early-morning warmth of a southern slope. For farmers with day jobs, who need evening light to get work done, a west slope will receive direct light longer. It would certainly be warmer on the west side, which is good in spring and fall, but not summer.

Flat land. Perfectly flat land is beloved for field and garden crops since it is so easy to work. It does not drain cold air well, however, which can be hard on fruit trees.

DRAINAGE

IF AN ACREAGE IS PERSISTENTLY WET because there is no lower level to drain water to, then forget about that property unless you're planning to raise cranberries or catfish. If it's wet because it was once drained and the old system has failed, then it's possible the acreage could be restored, though searching for and repairing old drain tile lines can be frustrating and expensive.

Signs of wet soils include wetland vegetation, low areas with signs of water pooling, and growth obviously stunted by flooding. The county Extension office or land conservation

IF A WET AREA in a field can't be drained, or was once drained and is now plugged up and wet again, consider (if you buy the land) preserving that area as a wetland. This adds to the biological richness of your farm, helping to satisfy the Organic Rule's biodiversity requirements. Often government funding is available to pay part or all of the costs of a wetland restoration. In this illustration, a water regulator prevents the restored wetland from overflowing and flooding the adjacent hay field.

department should have maps showing floodplains and high-water tables for the county; these are worth looking at if you're considering low-lying property.

FIELD LAYOUT AND USE

After you've noted a property's slopes and aspects, take a closer look at the shape of the fields and how the entire farm is laid out. Fields that are not square mean short rows at one or both ends, which means more time spent turning equipment and thus more fuel and time overall to get the job done. Fields that have poor access, are far from the buildings, or widely scattered from each other also decrease efficiency. If you have the time, however, then this type of farm may work for you. Inconvenient fields often mean a cheaper price for land in an agricultural area, since they are not attractive to large-scale farmers.

If it's hard to get a notion of the overall layout from walking the property, get an aerial photo from the county Extension agent, or online at the U.S. Geological Survey website at www.usgs.gov. Google Earth is even handier. Aerial photos can also reveal erosion and the type of vegetative cover.

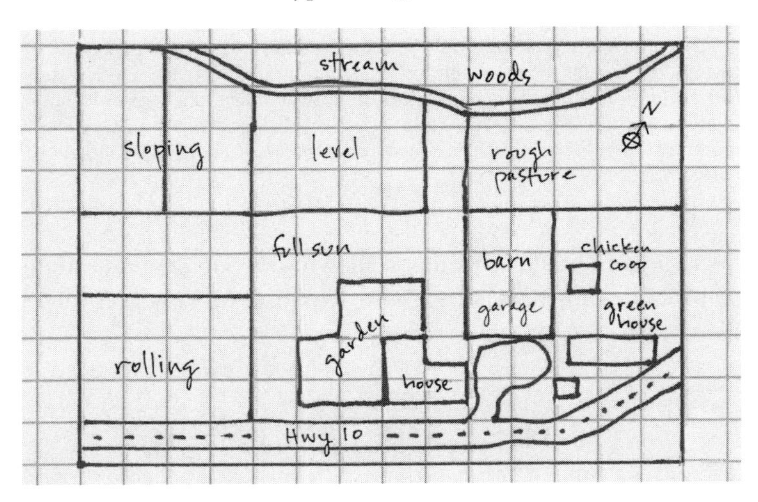

TAKE A PAD of graph paper and sketch a rough layout of the farm, noting types of land and significant features.

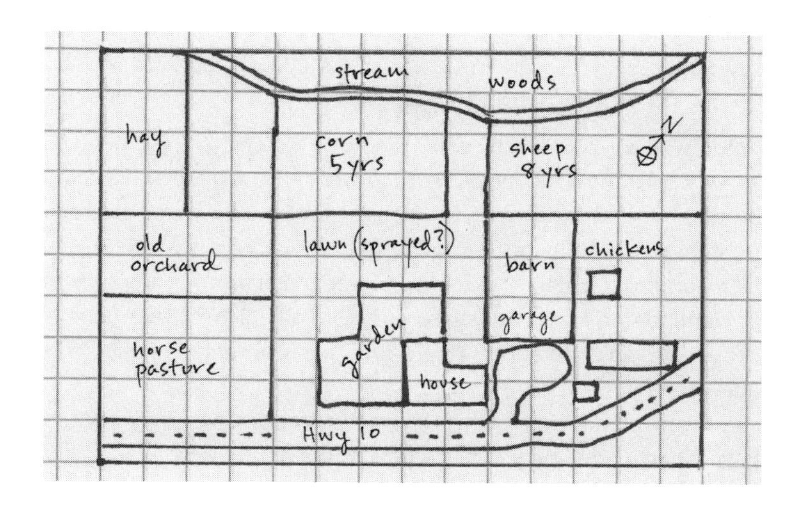

NOW FILL IN field histories for the past few years.

Field Histories

What has been planted in the fields and how many animals were grazed in the pastures in past years are good indicators of what a farm is suited for, and its ability to produce. Get a history for each field of what was planted, how much it yielded, what fertilizers, amendments, or chemicals were applied for at least the past three years, and further back if you can. Knowing how many animals the pastures supported is also useful.

The Organic Rule requires that no prohibited substances be used on land for three years in order for it to be eligible for certification. If you can determine that no prohibited fertilizers,

Surveys

When walking a property, walk the *entire* property. You'll need to know where the boundaries are, and what the neighbors are doing on their side of the fence. If they are growing conventional crops right up to the boundary, then you will be required, if you intend to be an organic farmer, to establish buffer zones on your property to prevent contamination of your crops from chemical drift or genetically modified pollen drift. This will reduce your acreage of certifiable crops.

If you intend to make an offer on the property, you may want to have a certified survey done as part of the deal. Since the original surveys were often made under primitive conditions and sometimes by less-than-qualified surveyors, a new survey frequently turns up some surprises — mostly little ones, sometimes major. Do not take a GPS out and do the survey yourself; this may satisfy you but will not stand up in the event of a legal dispute. Hire a licensed surveyor.

pesticides, and genetically modified seed have been used, and get a signed statement from the owner to that effect, the land may be immediately eligible for certification.

(Note: if you know what certification agency you will be using, download the "Prior Land Use Declaration" or equivalent document from the website and have the owner sign it.)

Current Use

Consider the current use of the property as well. Converting pastureland to crops is not that difficult. Converting cropland to pasture, however, requires a good "catch" (germination) of the seeding and enough patience for a sod to form that can hold up to hoof traffic. Both require hiring equipment, or investing in it yourself.

If the land is not being actively farmed, brush or trees may have moved in. Removing the unwanted vegetation will take quite a bit more work.

If the land is enrolled in a government program such as the Conservation Reserve Program (CRP), there will be some severe limitations on what you will be able to do until the contract expires. Examples include:

- no crop production
- no haying or grazing during primary nesting season as defined by the state Farm Service Agency (FSA)
- no haying or grazing at all unless authorized by your FSA office and then only for specified periods and no more than once every three years; or unless a drought or other natural disaster creates an emergency situation and FSA authorizes emergency haying or grazing

On the other hand, most CRP land is not sprayed with any chemicals, so if that land is coming out of a CRP contract and you verify that no prohibited chemicals or genetically modified seeds have been used for the past three years, you will be able to apply for organic certification immediately.

CHAPTER FOUR

BUILDINGS AND UTILITIES

Your home is important, my mom always said, and she meant that you should have a house that you find comfortable, convenient, and attractive.

If you decide you like the land and the buildings enough to consider making an offer, then do a careful assessment of what is wrong as well as what is right with the infrastructure. If there are major problems, get an estimate of what they will cost to fix. You'll need this estimate when you calculate your budget and figure out what to offer.

Help!

Assessing the worth and condition of a house is a big topic, with a plethora of resources. You may also wish to call in an expert. Here are some tips for whom to consult at which stage:

1. If you have a friend with some expertise, bring him or her along the second time you visit the property.

2. Hire a building inspector only at the point of making an offer.
3. If the inspector recommends further expert advice or you have some concerns, then bring in the relevant professional — septic, plumbing (these are two separate areas), electrician, HVAC (heating, ventilation, and air conditioning), roofer, and so on — for assessments and estimates.

I recommend hiring a building inspector for a house inspection, but assessing farm buildings, fences, and so on is simpler. You aren't dealing with so many hidden things (inside walls and ceilings) or with big systems — there's no furnace, minimal plumbing, and usually only basic wiring, so it's much easier to rely on visual inspections and do it yourself. If you truly feel incompetent to judge, find a farmer or Extension agent or even the bank's ag loan officer to come and walk the property with you.

Bare Land

If you plan to buy bare land and build your own home, *make sure* that the building site is suitable for a house, a septic system, and a driveway. Soil type, subsoil, groundwater levels, slope, climate, and whether you're living in an earthquake zone all impact what will be a good building site. It's a good idea to hire an engineer or building inspector to assess the site.

If you intend to build a nontraditional home or outbuildings, check online or with the county clerk to find which building codes are in force in the county to make sure you can get a permit if one is required — in most places, it will be.

If you're thinking about bringing in a mobile home, check county and local ordinances for siting requirements. If zoning is in force, mobile homes may not be allowed in the district, or only certain types of mobile homes.

If you're considering a nontraditional wastewater disposal system, such as an outhouse, a composting toilet, or a gray-water system, check the state and county department of health or sanitation websites for the regulations on septic or private wastewater disposal systems. Most counties will require permits to build a system. You should also ask the town clerk if there are township ordinances; some towns pass regulations that are more strict than those of the county (for more on land use regulation, see chapter 6).

IF YOU BUILD an alternative structure, such as a dome home, it's best to comply with building codes. Your new home will be safer and more salable as a result.

Many towns and counties are becoming less tolerant of alternative wastewater disposal systems; this is understandable in light of how groundwater contamination issues are cropping up in more and more communities. Even if you're allowed the composting toilet or outhouse, you still have all the water from the laundry, kitchen, and shower that has to go somewhere. Before you get too creative with your wastewater, find out what the rules are. Circumventing them may result in not only a hefty fine, but problems with your drinking water as well.

If you will be installing a septic system, have a "perc" (percolation) test done before making an offer, to be sure the soil is appropriate (instructions on how to get this done should be available from the county). If it is not, inquire about other options such as a mound system or simple tank. If no other options are available, you will have to look elsewhere for land.

If there is no driveway, make sure that you can get a permit for an access to a public road — whether it is owned by the township, county, or state. If you can't meet sight-line and distance requirements, you will not be able to build a driveway and may have to look for other land.

The House

If the house is ready to occupy, that's excellent, but many farmhouses will need some repairs or remodeling. This is not a big deal unless the work will significantly reduce the time or budget you need to get the farm operation under way. If the place needs expensive repairs, consider this when making an offer.

7 SIGNIFICANT REPAIRS AND HOW TO SPOT THEM

Here are the items that are priciest to fix and which, if not in good condition, will generally require immediate attention:

1. **The roof,** if it has shingles of the standard asphalt type, should be checked for curling at the edges, discolored areas, and cracked or missing pieces. If the roof is metal, look for rust (which paint should cure). Metal roofs should last forever if they are well cared for; the real worry is if water has been seeping through rusted or loose seams or nail holes, causing rotting in the underlayment or trusses.

2. **The foundation** should be square at the corners and not shifted, bulging, or cracked. If the house has a basement, look carefully at the basement walls for these problems. A shifted foundation may be evident in doors and windows that won't open or shut properly.

3. **Sill plates, studs, roof trusses, and other structural components** should be square and show no signs of rot, mold, termites, or carpenter ants. If necessary, have a professional pest inspection done. Certainly stick a pocket knife into any suspicious-looking wood (structural components are usually exposed in the basement and/or attic) — if it sinks in easily, there is probably dry rot and the wood will have to be replaced.

4. **The wiring** should be up to code. Old wiring can be a fire hazard and may also make the house uninsurable. Take a good look at the main electrical panel (usually

Inspection Time

Many (if not most) lenders may require a professional home inspection; even if it's not required, it's an excellent idea. You can ask your lender or real estate agent for names of inspectors, or visit the American Society of Home Inspectors at www.ashi.org to get names of inspectors in your area. If a home inspection finds things that are not up to code — such as a cesspool instead of a septic system — you may be required by the lender and/or the county to fix the problem. Ideally, you can get the seller to fix the problem before closing the sale.

in the basement or garage). An old box that still uses fuses instead of switches may be uninsurable. Also look to see if the wiring is neatly done. Rat's nests of wires, naked splices (two joined wires with no plastic cap), and similar sloppiness indicate a poor job, and possible problems elsewhere in the system.

5. **Plumbing** can have a variety of problems, beginning with old lead pipes or copper pipes with lead solder that can leach toxic lead into the water. Check also for water marks on ceilings and walls that may signal small leaks; these are sources of rot. If the water is very hard or alkaline, and the plumbing is old, there may be mineral buildup inside the pipes, which can slow water flow. At least replacing pipes is cheaper than replacing a well — usually.

6. **The furnace** may be hard to judge; some of the old ones will last for many decades, while many new models now have an expected lifespan of only about 15 years. Ask the age, maintenance history, and most recent maintenance date of the furnace, and if it seems prudent have a professional inspect it. Wood stoves and outdoor wood furnaces must have sound, well-maintained chimneys; chimney fires still destroy houses every year. Make sure that any wood stove installation is insurable: have an agent check it.

7. **Construction materials** may need immediate attention, such as old, lead-based paint in the interior, or asbestos insulation around pipes, an old furnace boiler, or in fiberboard form. This is fairly common in old buildings and a health risk only if the asbestos is disintegrating or disturbed.

Wastewater Disposal

Since it's a rare farm that is hooked up to a municipal sewer system, chances are that when you acquire a property you'll also acquire a septic system. It's best to acquire one that is not failing, since renovation or replacement of a system is a significant budget item. If the farm has a cesspool or dry well instead of a septic system, ask at the county offices if the county will require that a septic system be installed when the land has new owners.

SEPTIC SYSTEMS

In a standard septic system, wastewater flows into a septic tank, where the solid bits float or sink and are digested by bacteria. The fluid flows out the far end of the tank to the septic field, a series of subsurface, perforated pipes that let the liquid seep slowly back into the soil. Good soil will filter the water

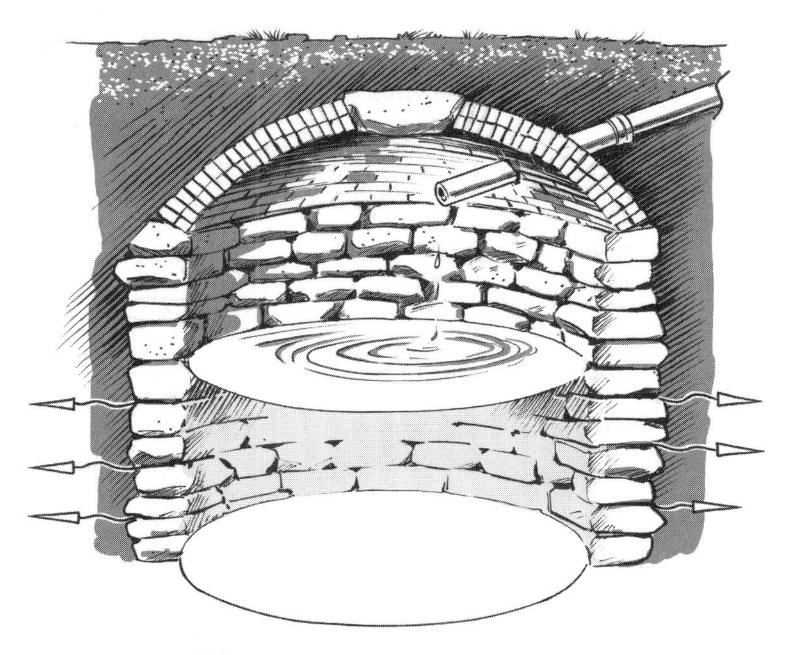

CESSPOOLS are still fairly common with old rural homes. They are frequently no longer allowed by wastewater disposal ordinances, and so will become illegal when the property changes hands. This issue should be addressed before closing on a property.

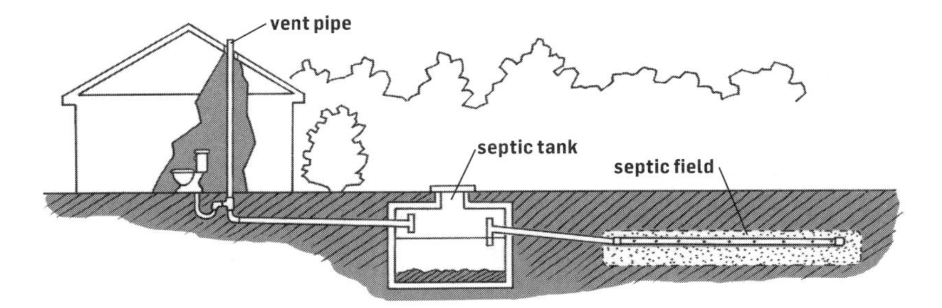

A STANDARD SEPTIC SYSTEM is a safe, effective method to dispose of household wastewater at no ongoing cost except having the tank checked and the solids removed occasionally.

clean; still, you don't want the septic system close to or uphill from the well.

Where the soil is inadequate for a septic field, the county may allow (at much greater expense to the homeowner) a mound system. This is simply piling the soil up high enough to create adequate filtering depth for the septic field.

If the soil is not even adequate for a mound system, then you may be allowed to install a holding tank (which is what we had to do), which you pay the local septic pumping service to empty on a regular basis and then take your sewage to the nearest small town's municipal plant. However, a small town with an aging sewage treatment plant may be reluctant to take on any more customers, and you could wind up with no legal, affordable options for wastewater disposal. For more information on wastewater disposal systems and regulations, check your state health department's website.

SIGNS OF SEPTIC PROBLEMS

A well-maintained septic system should last forever, but many are not well maintained. Signs of septic problems include:

- an odor of sewage in the basement or around the septic field
- mushy, soggy soil over the septic field
- slow toilet flushes

Sometimes the problem can be cured fairly easily by pumping and cleaning out the tank; but sometimes the septic field has become clogged with debris and will need to be replaced. If there are any red flags, contact the local septic service for an assessment of what the problem is, and what it will cost to fix.

Barn, Outbuildings, and Fences

Most farms come with an assortment of outbuildings, often less well maintained than the house.

Old barns are tough to let go, given the place they have in our culture and our hearts. But of all outbuildings they are usually the most expensive to restore, so do your calculations for this building with your head, not your feelings. If the barn is both decrepit and six times the size of anything you'll ever use, then get an estimate for having it torn down and either buried on-site or the debris hauled to a landfill. If the building is salvageable, however, visit the Barn Again website at www.preservationnation.org for resources and information on barn restoration.

Smaller outbuildings are simpler to tear down or fix up yourself, and, like barns, they can be "repurposed" or adapted to new uses. Corn cribs may be converted to wood ricks, an old

How to Look at an Outbuilding

For each outbuilding, note the condition of the foundation, the roof, and the structural components. Common problems include:

- Sway-backed roofs
- Dry rot in sill plates and studs
- Cracked cement slabs
- Rotting windows

Estimate what is past easy repair and should be torn down, and what could be fixed up and used.

granary turned into a guest house for your bed-and-breakfast, a pig sty turned into a creep feeder for calves. Consider the possibilities as you walk the property.

Look also at the overall situation of the buildings and their relationship to each other. If the prevailing wind means you always smell the barn in the house, this will affect your quality of life. If the farmstead sits at the bottom of a valley or slope, it may be very muddy around the buildings for long periods, which is a real pain. There should be enough turning and backing space around the buildings for livestock trailers, feed and seed delivery trucks, tractors and farm implements, and the septic pump truck.

FENCES

If the farm fences are decrepit, you may be able to run a temporary electric fence inside the old one to hold livestock (or keep the neighbor's livestock out) until you can get a permanent boundary fence rebuilt. Eventually, however, that job will have to be done, and it's a big one. Note also whether the type of fence is appropriate to the sort of livestock you want to raise: barbed wire is okay for cattle, but not a great idea for sheep. You should also familiarize yourself with fencing regulations, which are usually instituted at the county level, to see what part of a boundary fence you would be responsible for maintaining and whether a survey will be required before building or replacing a boundary fence on the property line.

FENCELINES and outbuildings on old farms are often decrepit; these can be taken out or rebuilt, if you have the time and budget.

Electricity

Though water and sewer are generally handled on-site in the country, electricity is delivered unless you are living (as most of us hope to someday) completely off the grid. If there is already service to the property, then you're all set. If you're looking at bare land and planning to hook up to the grid, then be sure to get an estimate from the power company of what it will cost to get a line in, since this can be unexpectedly expensive.

There's one further note for organic farmers. Power companies often keep the ground clear under their lines by spraying chemicals prohibited by the Organic Rule. If the property has power lines running across it, you should write to the power company each year requesting that no chemicals be used (this may be required by your certifier), and that a mowing or brush-clearing crew be sent instead. If you really want to make sure there is no spraying, you could keep the area under the lines cleared yourself with mowing or grazing.

Driveway

Most towns or townships and counties have ordinances requiring a driveway entrance to be a certain distance from road intersections and from the neighbor's driveway, and to have adequate sight lines so that drivers can see oncoming traffic and the oncoming traffic can see them. Sometimes a perfectly nice piece of property doesn't have enough road frontage, or it's not in the right spot, to meet driveway access regulations. At this point you can either walk away from that property, or see if an adjacent landowner will sell a strip of land for a driveway, or grant a legal easement for you to cross their land, and/or share their driveway (get the agreement in writing).

Some towns or townships and counties may also regulate the construction of the driveway itself, mandating minimum widths, adequate turning radius, ditching and culverts where necessary to maintain good access, and minimum roadbed standards. This ensures that emergency vehicles — fire trucks and ambulances — can get in and out without getting stuck.

In some situations a developer has broken a larger property into smaller acreages, building a private road to give all parcels access to a public road. This leaves the occupants of the development responsible for road clearing and maintenance. If you are looking at a rural development of this sort, don't believe the developer if he tells you the town or township will take over the road; contact the town or county clerk to make sure.

Other Hazards to Note

You never quite know what you might run across on a rural property; below is a list of a few other items that might (or might not) be a problem:

An old dump can be found somewhere on most farms, left from the days when there was nowhere to haul your garbage. ***What to do:*** Though unsightly, an old dump is not a hazard unless it was used for potentially hazardous substances. Poke around in the dump and see what's there, and whether it could affect the farm's water supply.

Leftover chemicals and old chemical containers could have been dumped into a sinkhole or stream, leading to water

contamination problems that can't be fixed easily. ***What to do:*** If you suspect this, have the water and stream bed soil tested specifically for ag chemicals and petroleum products.

Machinery graveyards also persist on many old farms. ***What to do:*** These are usually more a nuisance than a real hazard. They may also bring in a little cash when scrap metal prices rise, or contain something you might actually be able to use.

Underground storage tanks for petroleum products were fairly common on farms for many years, and you should ask if there ever was or still is one in the property. ***What to do:*** Leaks were also fairly common, so if there is a tank find out when it was last in use and have the water tested for petroleum products.

Old, half-forgotten fencelines may have long runs of rusty barbed wire hidden in the brush or just under the ground. ***What to do:*** These are dangerous for children and livestock and should be cleaned up, but in most cases shouldn't affect the sale.

An old unused well or two may be present and can be a hazard. ***What to do:*** These should be sealed and abandoned (see chapter 2) according to state guidelines.

CHAPTER FIVE

NEIGHBORS AND THE NEIGHBORHOOD

If buying a farm is like finding a spouse, then meeting the neighbors is like meeting the in-laws: ready or not, they're now part of your family. Fortunately, most rural residents are willing to lend an ear, lend equipment, and lend a hand. If you treat them just as well, you'll have a happy neighborhood.

LEARNING THE LAY OF THE LAND

LEARN ABOUT THE NEIGHBORHOOD before you make an offer on a property; it could save you a lot of regrets. Here are some valuable tools for finding out who the neighbors are and what they're doing on their land.

Google Earth. A terrific Internet tool, Google Earth is free (go to www.googleearth.com) and allows you to tour neighborhoods from the air, so you can spot the barns, mines, logging operations, and other land uses that from the ground may be hidden behind hills or trees. If you can't do this at home, go to the nearest public library with Internet service.

What you see on Google Earth can be correlated with the plat book (or county property records — see next page) to get the owner's name and contact information.

Paper road maps. Electronic navigation systems do not give you the big picture that a road map, topographical map, or DeLorme state atlas does of how roads, towns, government land, rivers, and lakes all fit together. You will find these at gas stations, county offices, tourism and economic development offices, local book stores, and through AAA (the American Automobile Association) and the U.S. Geological Survey (www.usgs.gov).

GIS

While you're nosing around on state and county websites, check what other information you can glean about the neighborhood. Geographic Information System (GIS) services, where available, make it possible to find all sorts of information about a neighborhood, from current land uses to depth to groundwater, floodplains, and many other useful maps.

Phone book. Stop by the local phone company and pick up a phone book, or check online yellow pages for local business listings, since many small-town businesses don't bother with websites. Other possibilities for getting a sense of what sort of businesses are operating in a town and area are the regional or county visitor's bureau, the local Chamber of Commerce, or Economic Development offices and their associated websites.

Plat books. Available in 12 states (Illinois, Indiana, Iowa, Kansas, Michigan, Minnesota, Missouri, Nebraska, North Dakota, Pennsylvania, South Dakota, and Wisconsin), plat books map all property lines in a county and include an index of owners. They are available at the county offices, from many real estate agencies, and sometimes in local bookstores (see http://farmandhomepublishers.com).

In most other states you can find out where property lines lie and who the owners are by asking at the town or county tax and/or assessor's office for the tax list and map, or checking websites. If "tax list" isn't the right phrase, try "parcel map," "landowner map," or similar phrases. Though paper copies of these documents are not for sale, you should be able to look at them at the office. In nondisclosure states — Alaska, Idaho, Kansas, Louisiana, Mississippi, Montana, New Mexico, North Dakota, Texas, Utah, Wyoming, and some counties in Missouri — none or only some property records are publicly available.

Nearby Farms

If there are large farms in the area where you're looking at property, don't let that scare you off before you research further. Size doesn't necessarily indicate quality. "Big Ag" has a bad name in many organic and sustainable circles, sometimes earned but sometimes not. Farms reflect their owners, and so are quite variable. If you're seriously interested in a property but worried about the neighbors, leave your opinions about conventional agriculture at home and go visit them to see what they're really like.

DON'T JUDGE A FARM by its size; not every small farm is well run, and not every large farm is offensive. A large-scale, grass-based dairy farm like this may make an excellent neighbor.

Red Flags Indicating a Problem Farm

- Lots of mud
- Silage plastic and other trash lying around
- Overflowing manure storage
- No vegetation buffers along streams
- No vegetation buffers around a barnyard or feedlot to filter manure runoff
- Crowded dirt cattle feedlots that continually generate dust and odor
- Crop rows run straight up and down slopes instead of across
- Gullies from erosion
- A casual attitude about mixing and applying chemicals

That said, there is normal farming and rural activity such as logging or gravel pits, and then there are truly objectionable things, such as a big farm upstream that has a problem with manure spills, or a noisy nearby sawmill. Some operations may seem much nicer than they really are: commercial orchards, for example, may be contaminating downstream groundwater with chemicals from spraying. Some types of neighbors and neighborhoods you may want to approach with caution, or avoid completely.

For example, a close friend and her husband drove to visit a property for sale in rural Oklahoma a few years ago. It seemed to have everything they wanted: pasture for their animals, some woods along a stream, and a bargain price. The land was beautiful, but the neighbors, as a kind retired sheriff told them, were mostly methamphetamine labs that kept a lot of mean dogs and did not cotton to strangers. My friends decided to look elsewhere.

LIVESTOCK OPERATIONS

I have been on large dairy operations where the cattle lanes were cemented, the manure covered, and surface waters carefully protected from runoff by buffer zones, and I've seen others that are mudholes. I've seen meat-animal finishing operations

TOTAL CONFINEMENT poultry operations can be identified by their characteristic long, low barns (though newer ones may be two-storied). Large operations may be hard to live near if you're downwind, but smaller farms that are well sited aren't usually offensively smelly except for a few times a year when it's time to clean out the barn(s). Commercial poultry feed contains arsenic, which has appeared in the groundwater in neighborhoods where there are a lot of poultry operations.

where the critters had constant access to pasture, and others where animals lived standing in their own manure. I have seen midscale poultry and pig operations where the barns and manure storage were sited and maintained specifically to minimize objectionable smells, and others where they weren't. Don't judge your neighbors or their farms until you've met them.

Also realize that crop and livestock farming is at times unavoidably smelly, messy, and noisy. Manure has to get spread, and crops have to be harvested before bad weather, even if it means running machinery far into the night. If you find these occasional nuisances intolerable, you should not be considering property in an agricultural area.

CHEMICAL AND POLLEN DRIFT

Cash-crop operations don't smell (unless they're spreading manure from a large livestock operation a couple of times a year), but you will get chemical drift if they spray on a breezy day. They may also use planes to spray (crop dusters); this increases the amount of drift and also the noise levels. Your organic crops will be decertified if they are contaminated by either synthetic chemicals or pollen from genetically modified crops. Depending on the crop and wind conditions, the pollen may blow a considerable distance.

This is an issue you'll need to discuss with an organic certification agency that knows the area. Organic farms deal with the potential for chemical and pollen drift by establishing buffer zones of trees, natural vegetation, or "sacrifice" crops along the property line wherever there is danger of drift. The width of the zone will vary according to the specifics of the site; this is determined by the certification agency.

Other Land Uses That Can Endanger Your Health

A lot more than farming goes on in the country. Land uses that can affect your water quality are the ones you especially want to check out; the most likely of these to cause problems are mining for coal, gas, oil, and metals.

COAL MINING

Coal mining is most familiar in the Appalachian Mountains, but it also occurs in areas of the Midwest, Texas, Alabama, and the Rocky Mountains, according to Curtis Seltzer's book, *How to Be a Dirt-Smart Buyer of Country Property* (see Recommended Reading). Underground coal mines, Seltzer writes, whether

Stray Voltage

Stray voltage occurs when the ground return on power lines is inadequate to handle the flow of electricity, so that current escapes into the earth and turns up in places like the watering cups in the dairy barn. Farmers in several states have filed lawsuits against power companies claiming illness and injury caused by stray voltage; the power companies have denied responsibility.

Finding out if there are stray voltage problems in a neighborhood may take some digging. Ask your Extension agent, and visit www.strayvoltage.org for background information.

active or closed, can cause surface subsidence, acid drainage into surface water, and other problems, including long-burning underground fires. Surface mining is not only ugly, but it can also cause flooding from increased runoff when mine establishment sacrifices the original vegetation.

Finding out the future development plans for more mines in your area can be difficult, since companies are secretive about their plans. If you are concerned, start by asking at the state

ABANDONED COAL MINE SHAFTS are a double red flag: they are a hazard to curious people (and animals) and may indicate deeper problems with groundwater contamination.

and county levels what permits have been issued for mining and drilling in the area, and ask if any applications have been filed.

HYDROFRACKING

"Frac" mining for natural gas has exploded in recent years, especially in the 575-mile-long (925 km) Marcellus Shale formation, 8,000 feet (2,438 m) below the surface of the Appalachian Basin in New York and Pennsylvania, and is causing major problems with water quality, noise, and aesthetics for some area residents. For an excellent summary of the frac mining process and its consequences, as well as ongoing coverage of the issue, visit www.propublica.org (click on "our investigations" in the menu bar).

Frac mining presents similar issues as mining for metals in dealing with toxic chemicals seeping into surface and ground

High-Capacity Wells

A high-capacity well by definition delivers more than 70 gallons (265 L) per minute, and enough of them in an area can lower water tables to the point where other nearby wells run dry. Large irrigated crop farms and large dairy operations are the types of farms that commonly have such wells. The county clerk should have records of high-capacity wells operating in the area and if any are currently in the permitting process; also check with the state department of natural resources to see if groundwater levels are dropping in the area.

water; metals mining, however, is much more tightly regulated than the relatively new sector of frac mining. Contact your state department of natural resources and ask what mines are operating in your area. Editors at the local or regional newspaper, radio, and TV stations will know if there have been any problems or controversy, or check the archives of regional and local news outlets for stories on frac mining.

OTHER EXCAVATIONS

Small-scale gravel pits, rock quarries, and sand mines may not be the prettiest things in the world, but they are fairly benign in environmental terms. Old quarries and pits may also furnish some local recreation by way of swimming holes and backstops for target practice.

On the other hand, large-scale sand, rock, and gravel mining alters the landscape and creates a lot of truck traffic, noise, and dust. Those operations should be avoided as neighbors.

Land Uses You May Simply Find Obnoxious

Though we have a cultural image of the countryside as being quiet, with dark nights and wonderful clean air, that is not really true in some rural areas. When considering a property, try to visit it during both weekdays and weekends, after dark, and in different seasons. For example, if you were never in our neighborhood in winter you wouldn't know about the snowmobile traffic. In other areas, dirt bikes and ATVs might be a real annoyance on weekends. Before you purchase property, consider other situations that you might not want to live with:

Poor air quality due to nearby light industry; or, in winter, smoke from wood stoves in a low area where it can hang for long periods.

Excessive noise from sawmills, gravel pits, public shooting ranges, racetracks, outdoor bands at the campground across the lake, the flight path for commercial airliners or a military flyover zone, or a heavily used highway.

Lights from the barns and yards at nearby farms, headlights sweeping across the house windows due to a curve in the road, or glaring lights from other nearby businesses or neighbors.

Trespassers

Trespassing is fairly common in many rural areas, and can be benign, annoying, or dangerous depending on the offenders and their purposes. Look for cut fences, four-wheeling trails not used by the owners, or a path to the spring coming from the wrong direction.

A common problem for properties on back roads is illegal dumping, especially of items that may be charged for if you take them to the official dump or recycling center, such as tires, mattresses, and old appliances. Finding such items in the roadside ditches or brush signals a problem. If the property has woods, you may also be dealing with trespassing hunters, and possibly with timber theft.

ATV AND SNOWMOBILE TRAILS are common in rural areas; if you don't enjoy the roar of the machinery (especially at night), make sure there aren't any trails near the property you're considering. Of course, if you love to ride, you'll want to be as close to trailside as possible.

Poor visual aesthetics, due to highly visible cell phone and radio towers, billboards, or other things you don't want to look at every day.

Dumps and landfills in the area, especially if there is a lot of truck traffic as a result.

FUTURE TRENDS

There is one last item to research when checking out a neighborhood. Is the area growing, dwindling, or static? The answer to this question may determine whether your farm can exist in the long run.

Residential Sprawl

Since rural areas are by definition primarily open land, development can take almost any form and sometimes does. The most common is residential development, where fields become housing subdivisions. This tends to bring with it everything from leash laws and complaints about farms that run machinery too late at night to dropping groundwater tables from too many wells and nitrate contamination from too many septic systems too close together. A livestock operation may find it increasingly difficult to exist in the face of this type of development, while fruit and vegetable growers may thrive as their potential customer base grows.

Where to learn more. The quickest way to get a sense of how fast and where this type of development is taking place is to find out at the county offices how many well permits have been issued in the county each year for the past 10 years, or the number of driveway access permits or building permits.

Farm Expansion

In areas of prime agricultural land, big farms are usually getting bigger. This can cripple nearby small towns, as the overall number of farmers dwindle, and the ones who are left are big enough to buy their feed, seed, fertilizer, and chemicals wholesale rather than from the local dealer. The drain of money and people forces retail stores to close and schools to consolidate.

An influx of small farmers selling locally produced food may be just the ticket to reverse the trend, but it is a long uphill battle. Though the cheap housing and long distances to bigger towns may be attractive, think long and hard about how you could

IN AREAS WHERE large-scale has replaced small-scale agriculture, many small towns suffer from population loss and economic hard times.

survive economically in such an area before deciding to take the plunge, especially if you are the only one bucking the trend.

Where to learn more. Ask the county Extension agent what the trend is in farm sizes, and check your state's agricultural statistics at the state department of agriculture, or the National Agricultural Statistics Service at www.nass.usda.gov. (For more about future land use trends, see chapter 6.)

Industrial Operations in Nonzoned Areas

Counties and towns or townships with no zoning laws are very vulnerable to undesirable land uses that are zoned out of other areas. These include landfills, large livestock operations (called CAFOs, for concentrated animal feeding operations), and industrial plants that need lots of water and/or generate a lot of light, noise, and smell. While zoning necessarily restricts what you can do with your property, it also protects you from the neighbors doing something awful with theirs. (For more on government regulation, see chapter 6.)

Where to learn more. Ask at the county or town offices if zoning is in force, and if it is, get a copy of the ordinance and the map of the zoning districts. These will tell you precisely what land uses are allowed or prohibited in which districts, and where those districts are.

Once again, it's also an excellent idea to subscribe to area newspapers — especially the ag papers — and listen to the local radio and TV news. These are often the first to let people know what changes are in the wind.

CHAPTER SIX

GOVERNMENT REGULATIONS AND SERVICES

The four levels of government — federal, state, county, and town or township — all have some say in how you will be able to use your land. Most often, though, rural property owners deal with government at the county and township levels regarding land-use issues.

Different levels of government also are responsible for enforcing building codes; road construction and maintenance; mail delivery; fire, police, and ambulance services; schools; and libraries. For all these services, you pay property taxes, often at a reduced rate if you meet the criteria for being an active farm and your state has a property tax reduction program for farmers.

When you've narrowed your property search to a particular county or counties, one of the first orders of business is to visit the county website or drop by the county offices (usually in the courthouse building) to learn how the government regulations specific to that area might impact your farm plans.

LOOKING INTO THE LEGALITIES

At the county or town level, take a look at the following:

All ordinances regulating land use, especially any zoning ordinances, and the map of zoning districts.

Ordinances that state the standards and permit requirements for building a home, constructing a driveway, constructing a driveway access to a public road, drilling a well, and installing a septic system.

Permits may also be required for such things as on-farm sales or some other types of on-site businesses, such as a machinery repair shop or a campground.

The comprehensive plan for the county, if there is one. This will be an intimidating document, but you don't have to read the whole thing. You should pay special attention to the sections on transportation and utilities, to see what future roads and power lines are planned, and the sections on land use, agriculture, and economic development.

After you've received this basic information from the county, call the town chair or town clerk (find the name through the county), and ask what permits the town might require in addition to those required by the county (or check online to see if there is a town website, or a link to town-level regulations from the county website). Also ask if there are any township-level land-use ordinances. If the county zoning map doesn't show zoning for this township, ask the town clerk if town zoning is in force, or, if it's not, if zoning is being discussed.

BUILDING CODES

Frequently, building codes aren't enforced until a property changes hands. If a house is already on the land, along with a well, driveway, and septic system, do yourself a favor and get a professional inspection to make sure everything is up to local codes before you buy. If it's not, negotiate with the seller to cover the cost of getting it up to code. You don't want to buy a house only to find that you have a cesspool instead of a septic field, and must install an expensive new septic system.

If you are looking at bare land with a dream to build something other than the usual stick-built home, don't despair of getting a permit. Most building codes in the United States are based on the Uniform Building Code established by the International Conference of Building Officials (ICBO). The ICBO has also developed approved designs for domes, yurts, and some other common nontraditional structures. Work with your local building inspector or contractor to obtain the right design criteria.

Land-Use Regulation and Zoning

Because different land uses don't always make good neighbors — such as subdivisions and landfills, or large livestock operations and campgrounds — counties and towns may establish zoning ordinances to regulate what can happen where. This preserves the current residents' quality of life and limits conflicts; of course, it also limits what can be done on your land. In townships where zoning is not in force, other types of ordinances may be used to at least limit the effects of objectionable activities.

3 TYPES OF ORDINANCE

Counties and townships use three types of ordinances to regulate land use and development: licensing and nuisance ordinances, land division ordinances, and zoning ordinances.

Licensing and Nuisance Ordinances. These regulate activities that can be harmful or offensive by setting rules that must be followed in order for that activity to occur. The town or county must grant an operator's license to those who follow the rules. This type of ordinance can't regulate either the type of land use or where it occurs, only such things as hours of operation or pollution levels. Example: if an adult entertainment center wishes to build in your township, this type of ordinance can regulate its hours and lighting (within reason), but only if the ordinance is in place before the establishment is built. This type of ordinance cannot stop the construction or operation.

Land Division Ordinances. These regulate how land can be divided in preparation for development. These rules make sure a parcel is buildable, has road access, and has an acceptable wastewater disposal system. Like licensing and nuisance ordinances, land division ordinances can't regulate where or what kind of development can occur, but they can establish minimum lot sizes and how many residences may occupy a parcel.

Minimum lot sizes are common in many rural areas; if you're planning to buy a parcel and then sell half to cover your costs, this type of ordinance may prevent such a plan. If you're planning to buy land with several other people and then all build your own residences, this type of ordinance will determine if it's possible on a particular piece of property.

Zoning Ordinances. These regulate both the kind of land use that occurs and where. Zoning offers maximum protection against your neighbor doing something you don't like; it also offers the maximum limitations on what you can do on your own property. Example: in a residential-only zone, your neighbor may be

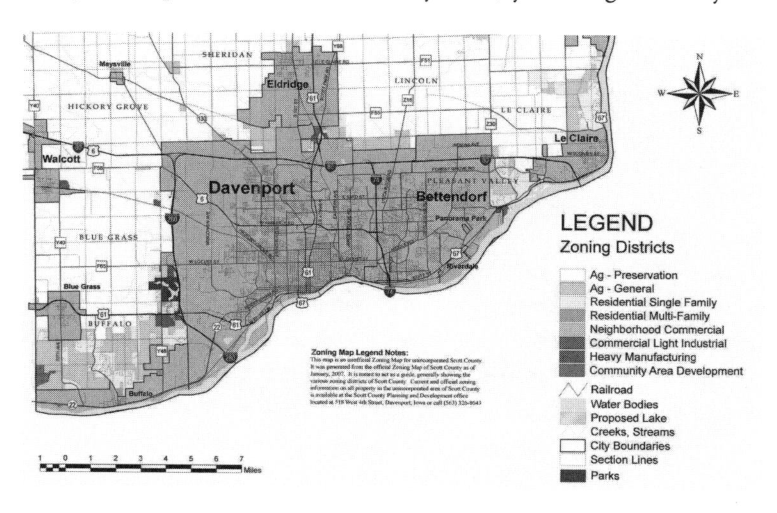

A ZONING MAP shows zoning district types and boundaries. The accompanying zoning ordinance details what land uses are allowed, prohibited, or must have a conditional use permit in each district.

prohibited from opening a gravel pit next to the property line, but you may be prohibited from opening a vegetable stand.

The more sparsely populated the area, the less likely it is to have zoning. Unfortunately, this makes these areas targets for some of the land uses — such as CAFOs, demolition landfills, racetracks, and subdivisions — that many rural residents find objectionable.

On the flip side, zoning may limit your ability to keep livestock or spread manure, operate an on-farm business, live in a mobile home, or cut your own trees for firewood or timber. It all depends on the details of the zoning ordinance.

If zoning is in force in your target county, get a copy of the ordinance and the zoning map. The map shows what the zoning districts are, and where they are. The ordinance will tell you what uses are allowed, prohibited, or conditional in each district.

Consider all the things that you might now or someday want to do on your farm, and see if the ordinance will permit it. This might include such things as operating an on-farm retail store, running a bed-and-breakfast, teaching homesteading classes, or managing a used farm equipment lot.

If there's no zoning in force, consider what sorts of development the area might be targeted for, including prisons, adult entertainment establishments (this has been a real problem in some townships in my area), and large-scale agriculture. I've yet to meet anyone who enjoys living next to a big swine operation.

STATE REGULATIONS

Lastly, because some land uses — such as CAFOs and metallic mining — are so objectionable to many people, the state

What Lies Beneath

Sometimes the mineral rights for a piece of property are owned by someone different than the owner of the land, and if that's true of your property, then it is always a possibility that a mining contractor will show up some nice morning to begin operations. Though the possibility is remote in most areas, if the deed to the land doesn't include mineral rights, seek legal advice.

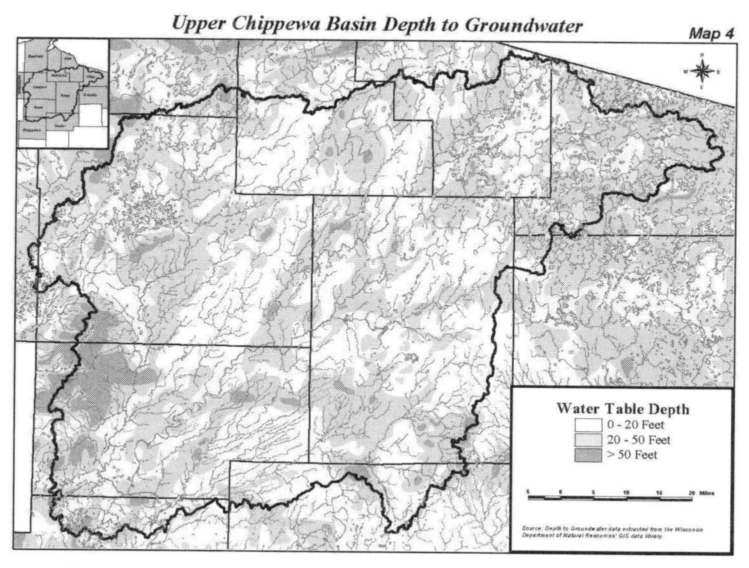

COMPREHENSIVE PLANS and other documents usually have a lot of useful information on towns. For example, this "depth to groundwater" map is helpful if you hope to install a septic system, have a dry basement, or want to avoid getting the tractor stuck when you do your spring tillage.

may have taken regulatory power away from counties and towns for that particular use and imposed its own permitting process. If the proposed use meets the state requirements, it can locate anywhere in the state, despite any outcry from local residents.

There is one limited exception, however. If a town or county has zoning that prohibits that use except in a designated district where it *is* permitted, then the town or county may (depending on how the state regulation is set up) have some say in where that use can take place.

Property Taxes

Property taxes are what pay for many government services and are assessed according to a property's fair market value (FMV) or a set percentage of FMV, which can be quite high. Fortunately, many states have tax breaks in place for farmers, generally referred to as "use value" taxation. This means that if you actively farm the land, it will be taxed at farmland rates, not FMV. In some cases, you may have to meet criteria for being an active farmer in order to qualify for the tax break.

Ask the following two questions at the county offices:

- At what rate have property taxes risen over the past 10 years?
- What programs are in place to protect farmers from rising taxes?

The answers will indicate whether you'll be able to afford to farm in years to come. For example, our property taxes have more than tripled in 20 years, and would be higher yet if not for use-value taxation.

Protections for Farmers

"Right to Farm" laws protect farmers from being prosecuted for normal farming activities that new neighbors find to be a nuisance. These activities usually include manure spreading, running machinery very early or late, and other normal odors, lights, and noise associated with farming.

Depending on the state, these types of laws may be passed at the state, county, or town level; the county administrators should know what regulations are in effect.

If you are looking at land in a more densely populated area, and planning on a type of farming that involves noise, smells, or lights that may not fit well into the neighborhood, check on what Right to Farm laws are in place. Since these laws were written primarily to protect existing farms from the complaints of new neighbors, be especially careful to ask if the law will preserve your right to farm when you are the newcomer in the neighborhood. It might not.

Environmental Restrictions and Conservation Programs

Land near wetlands, shorelands, the habitats of endangered or threatened species, and natural heritage sites will have some restrictions on what can be built and where, and what farming activities will be allowed. For example, manure spreading may be prohibited within a certain distance from surface waters, or allowed only if you have an approved "nutrient management plan" or similarly named document on file with the appropriate government office. Your best sources for this type of information are the county Extension agent and the nearest branch office of your state's department of natural resources.

If the land you're considering is enrolled in a state or federal conservation program — the most commonly used one is the Conservation Reserve Program (CRP) — then that land can't be farmed until the program contract expires. On the other hand, CRP land usually hasn't had any chemicals applied or GMO seed used (though you must check on this) since the beginning of the contract, and so may be immediately eligible for organic certification.

GOVERNMENT SERVICES

As well as regulating your activities, your town, county, and state connect you to your community through various services.

Roads

In rural areas there are state highways, county roads, and town roads, all color-coded as such on most road maps. Questions, complaints, and compliments about road maintenance, repair, and plowing should be directed to the appropriate level of government. You'll want to ask the neighbors, too, how often the road to the land you're considering is impassable due to flood, snow, or other natural emergencies. A truly epic weather disaster may close down every road in the county once in a while; what you don't want is a road that regularly washes out or becomes otherwise impassable. This is especially true if you have perishable farm products you need to get to market.

Private roads are built by developers or property owners and maintained by property owners. A common complaint of land buyers is that the developer told them the private road would be

IF YOU NEED TO GET PRODUCTS to market, kids to school, or yourself to a job, you'll need to live on a well-built and well-maintained road.

taken over by the town and thus the property owners would not have the hassle and expense of maintenance.

This is usually not the truth. In general, towns won't consider taking over a private road unless it's been built to the town or county's road construction standards, which tends to be expensive. Secondly, many if not most rural towns are already stretched to provide services, and they aren't looking for more roads to maintain. In our township, it is now standard policy to decline taking over any private roads, no matter how high quality the construction.

BUILDING YOUR OWN ROAD

If you're buying bare land and putting in the access road yourself, first get the road specifications from the county. If your road is not built up to the minimum standards, the county may not issue you a building permit when it's time to put up the house. You may also need a permit to connect your road to a public road.

Other Government Services

Since people are thin on the ground in rural areas, emergency services — fire, police, and ambulance — tend to be thinly spread as well. Police coverage is usually provided through the county sheriff's department, while fire and ambulance are more often provided by the town, by one town contracting with another for the service, or through special fire and ambulance taxing districts. Because distances are big, and volunteers may first have to travel from their homes to the emergency vehicles, response times to emergencies tend to be considerably longer in rural areas.

For this reason, property insurance rates are usually quite a bit higher in rural areas, and a house fire often means a total loss. And if you have a medical condition that may require a fast response time, you should carefully consider where you're planning to locate.

On the plus side, you'll get free mail delivery just about anywhere, though only to the far end of your driveway. School bus service may be a little more difficult; sometimes parents may have to drive their kids to a bus stop a mile or two down the road. Call the local school district to find out.

CHAPTER SEVEN

FINANCING A FARM

No one gives land away these days, so you must have cash. Since almost no one has enough cash to buy land outright, most folks have to find someone who will lend them the cash.

Money Down

Nearly all lenders also require a down payment of a certain percentage of the purchase price (the exact percentage will depend on the lender and varies widely), so that you have some serious skin in the game and have less incentive to walk away if you can't make the mortgage payments.

You may, with planning, persistence, creativity, help from a seller, and perhaps from some sort of beginning farmer assistance program, be able to acquire land with little or nothing down and minimal payments. Every beginning farmer should explore nontraditional possibilities and resources for getting onto land because, frankly, farm loans are tough to get, even if you have a farm business plan, a down payment, and an income separate from the farm. For further discussion of these options, see chapter 8. (Keep in mind that many farmers never own much land; they lease or rent. This chapter, however, focuses on purchasing land using traditional financing.)

You will still need some cash. Even if you are able to buy with no money down, you will still need to make payments, and you'll need start-up cash to buy seed, livestock, equipment, and infrastructure, as well as clothes and groceries, and to pay the property tax bill, which sadly will arrive every year and is separate from the mortgage.

The first thing you must do to get land, therefore, is to save up enough cash, or get a reliable job, or preferably both, so that you can make the mortgage payments and cover start-up expenses.

The second thing you must do is find a potential lender and convince them that you're a good risk.

LOANS AND THE BEGINNING FARMER

LENDERS DECIDE WHETHER or not to give you a loan based on statistics, what they know about your business, and their judgment of your ability to plan, manage, and be realistic in your expectations. This is how they assess the likelihood of having the loan paid off on time and in full. If they believe that likelihood is low, they won't make the loan.

Statistics. The odds are against new farmers, and especially against new small-scale sustainable farmers. According to the USDA, just 22 percent of beginning farmers turn a profit in their first year.

Insufficient farm income. In common with all agriculture, most organic farmers rely on income from off-farm jobs. Just 14.3 percent made their entire income from the farm in 2007, and 57 percent made less than 25 percent of total household income from the sale of farm products. For this reason, a plan to make all your income from the farm is unlikely to impress a lender unless backed by detailed planning and extensive experience.

Uncertain markets. Though the national resurgence of interest in locally produced foods has been wonderful for farmers who sell directly to consumers, many farm lenders are not yet comfortable making a loan whose payback depends on profitable local markets for the foreseeable future. A loan officer has to justify a loan decision with data; there may not be enough long-term data on organic and local markets for an officer to make a convincing case to superiors that a loan should be granted.

Lack of business experience. The Land Stewardship Project's 2003 survey of ag lenders found that important additional reasons for denying loans were a lack of business/marketing plans and poor management skills on the part of the applicant.

For these reasons — and this could be the strategy that makes you eligible for a loan — consider NOT applying for a traditional farm loan, but applying for a loan simply for the land and house. This is possible only if you or your partner have an off-farm job that you or they intend to keep. Income from the farm operation is not even considered in calculating whether you can make the mortgage payments, and this is a much more comfortable scenario for most lenders.

How to Be a Good Credit Risk

A real eye-opener for beginning farmers was the Land Stewardship Project's June 2003 "Farmer Lender Educator Survey on Sustainable Agriculture and Credit." The survey asked lenders what made them look favorably on a mortgage loan application.

Most lenders wanted:

- Three or more years of financial records
- A detailed business plan
- The amount of equity
- The applicant's background, experience, and management skills
- Total household income

Therefore, in order to make a proposed farm business look good to a lender, you will need:

- A detailed business plan
- Some equity — that is, cash to invest in the land and the business
- Some solid experience
- A good credit score
- Off-farm income, at least through the start-up phase

So, if you don't have experience or cash or a business plan or any of the three, then you should consider getting an internship or farm job, saving some money, and taking a class or enlisting other help to develop a business plan. There is plenty of help available online, at your public library, or through various federal, county, and state agencies.

None of this will happen overnight, or probably even within the next year. This will take patience, perseverance, planning, and self-discipline, all excellent traits that will be very helpful when you get your land and start farming. A great place to begin researching funding sources and loan requirements, as well as assistance and resources for developing a business plan, is the "Small Farm Funding Resources" page of the USDA's National Agricultural Library's Rural Information Center website at www.nal.usda.gov.

POTENTIAL LENDERS

IF YOU RANK POTENTIAL LENDERS according to how easily they grant loans, and how good a deal they offer on the size of the down payment and interest rate, then generally family will

Look for Less-Than-Perfect Land

Prime farmland is expensive, so look instead at the bits and pieces, the less-than-perfect land that no large-scale farmer wants. Small-scale operations can fit into a lot of the nooks and crannies in the landscape that aren't priced so high.

come first, followed by the seller of the land and government programs, and lastly commercial lenders.

Family or Private Individuals

If your family has money to spare and is willing to lend it to you so that you can buy land, you are extremely fortunate. Most often, parents can't finance the whole purchase but some are able to help with the down payment. To keep family relations amicable, draw up a legal document specifying the amount of the loan, the interest rate (if any), and the payment schedule.

Another possibility — admittedly remote — is an unrelated private individual, who has money and is willing to take a chance on being your silent partner. To prevent tax or legal complications down the line, have any agreement you make drawn up by an attorney.

The Selling Landowner

Land contracts, also known as a deed of trust, installment contract, or contract for deed, are fairly common in rural areas depending on state laws and the details of the contract.

With a land contract, the landowner becomes your lender. The buyer usually makes a cash down payment, and then pays off the rest of the price directly to the landowner in monthly, quarterly, or annual payments until the debt is retired.

For the buyer, the biggest drawback to this arrangement is if you fall behind or fail to make payments. In most cases, the land then reverts to the original owner and you lose not only the land but all the money you've already invested.

If you decide to pursue a land contract, *it's essential to hire a lawyer* who understands real estate to help you work out the details and draw up the contract. This protects both parties from the many pitfalls that can trouble a land purchase.

Government Agencies

Since getting a loan to buy farmland has been such a chronic difficulty for new farmers, the government stepped in beginning in the 1930s with programs to help ease the way. There are two types of government assistance: outright loans and government-backed loans from commercial lenders — that is, the government guarantees that a loan from a bank will get paid off, even if you default. This makes participating banks much more willing to lend money to new farmers.

The alphabet soup of government agencies can be confusing to say the least. At the national level, there are three programs that offer loans to new farmers for buying land.

The Farm Credit System (FCS). The Farm Credit System is often called the Federal Land Bank; in fact, the Land Bank is just one of the three banking divisions of the FCS, but it is the one that makes loans for farmland. Although the FCS was organized by the federal government, local farm credit associations are actually cooperatives owned by the folks who borrow from them. If you receive an FCS loan, you will also be required to purchase stock in your local FCS branch.

The FCS deals primarily in farm loans but also makes rural housing loans. Having an outside income helps a lot when applying for FCS loans. For more information, go to the FCS website at www.fca.gov/FCS-Institutions.htm.

The Farm Service Agency (FSA). Part of the USDA, the FSA can make or guarantee short-term loans for land, assuming that once the farm is up and running you will qualify for commercial credit. This means a borrower has to make a convincing case that your farm will be profitable. You must first be refused by commercial lenders before you can apply for an FSA loan guarantee, and if you are refused even with the guarantee, then you can apply for an FSA loan. For more information on FSA farm loan programs, go to http://www.fsa.usda.gov/FSA/webapp?area=home&subject=fmlp&topic=landing.

USDA Rural Development. Technically, the RD does not offer loans for farms, but it does offer both direct loans and guarantees for rural housing loans. This may be an option if you qualify for assistance and do not plan to make any income from a farm operation. For more information, go to www.rurdev.usda.gov.

Two other government agencies don't make direct loans, but may guarantee loans: the Federal Housing Administration, and the Veterans Administration. Obviously you have to be a veteran to apply for VA assistance.

Check at the state level for programs and assistance for new farmers. Start with the state's agriculture and economic development department websites. Or you could do a subject search, using the terms "rural development," "farming," "mortgage," and "Beginning Farmer Loan Program."

Banks, Credit Unions, Savings and Loans

Perhaps your family doesn't have the money to lend, the seller needs the cash now, and there isn't enough funding or you don't qualify for government programs. In that case, your only option for a conventional mortgage is a bank, or its close relatives, credit unions and savings and loan associations. Your best chance of getting a loan is from a *locally owned* bank, whose officers know the farmland in the area and the local farm economy, and will take the time to get to know you. Having off-farm income is usually essential to getting a loan.

If you schedule a meeting with a bank's ag loan officer, be prepared, because it is essential to make a positive impression: the officer at a local bank can make your character a part of the decision of whether or not to provide a loan. When you go in for an appointment, bring along as many documents and credentials as possible. With luck and good preparation, it may be the beginning of a long and beneficial relationship.

CHAPTER EIGHT

ALTERNATIVE WAYS TO GET ONTO LAND

There are other ways to structure a land purchase, and ways you can farm even if you don't own land. You'll still need some money, but perhaps not so much up front — or not as much so soon — as if you went a more traditional route.

3 First Steps

Begin with these basic steps, in whatever order you prefer, or all at once if possible!

1. **Gain knowledge and skills.** If you intend to farm for a business, then get the farm and business planning experience that will prepare you to run your own operation. This way, when opportunity knocks, you'll be ready.

2. **Research the many programs** designed to assist new farmers in gaining access to farmland. We'll discuss here some national and a few regional organizations involved in this issue; you should continue your search on the Internet by focusing on local, state, and regional organizations and by talking to other farmers and educators. An excellent place to begin is www.start2farm.gov, a national clearinghouse for new farmer programs.
3. **Consider every possible nontraditional option** for actually getting onto land. None of these alternative concepts is difficult to understand, but you will need to consider the advantages and disadvantages of each, and if you pursue one of them, get an attorney involved to protect all parties.

A HANDS-ON EDUCATION

You are going to ask educators, farmers, and sustainable agriculture organizations for their time, expertise, and financial help to assist you in obtaining a farm. What they want out of the collaboration is a successful new small-scale farm operation. Your part of the deal is to get the hands-on farm experience and business expertise that will make it possible to successfully run and manage such an operation.

Farm Jobs

If you didn't grow up living or working on a farm, your options for gaining hands-on experience are to find an apprenticeship or internship or simply a job on a farm of the type that you hope to one day operate. Farm jobs tend to be advertised locally, since that's where potential workers live. Check Craigslist (www.craigslist.org) and other print and electronic "help wanted" ads. Or stop by a likely looking farm and ask if they're looking for help.

If you don't need to get paid, check out the Worldwide Opportunities on Organic Farms website at wwoof.org as well as its North American chapter at www.wwoofusa.org. Host farms offer room and board in exchange for work; the goal is a learning experience that is both cultural and practical for the volunteer.

Farm Apprenticeships and Internships

Apprenticeships and internships usually offer room, board, and some pay (though probably less than a straightforward farm job) and, with a conscientious farmer, additional education by way of such tools as reading assignments, visits to other farms, attendance at area events, and field days.

If you apply for one of these positions, do so early since many are quite competitive. Since not all internships are listed on Internet databases, ask around at local farmers' markets, field days, and sustainable ag conferences where other farmers gather.

Many internships are listed on the Internet. Some good places to start are:

The National Center for Appropriate Technology's National Sustainable Agriculture Information Service (formerly ATTRA — Appropriate Technology Transfer for Rural Areas) offers probably the most comprehensive available list of internships and apprenticeships. Visit the website at www.attra.ncat.org and click on "Education." While there, also check out the "Beginning Farmers" page for other assistance programs for new farmers.

Collaborative Regional Alliance for Farmer Training (CRAFT) began in 1994 in the Hudson River Valley region and is still going strong. These well-structured programs involve a number of farms in an area cooperating to give their interns a broader experience than they would obtain interning on a single farm. Typically, the programs include classroom work, regular tours of other farms, and the opportunity for interns to network with one another. To find if there is a CRAFT program organized in your state or region, check the website at www.craftfarmapprentice.com. It provides links to other CRAFTs and other similar programs.

Good Food Jobs at www.goodfoodjobs.com lists a mix of apprenticeships and positions requiring more advanced skills, some on the farm and others in related sectors such as farmers'

FARM INTERNSHIPS, apprenticeships, and plain old farm jobs can be the best way to learn how to farm — and whether you truly love it.

market managers, local marketing of farm products, and sustainable ag education. If you're past the beginning intern phase and looking for a job that keeps you in the sector but pays a little better, take a look at these listings.

Many state-level organic and/or sustainable farming organizations list apprenticeship and internship opportunities on their websites, as do some of the Small Farm Institutes associated with land grant universities. Find out what's available in your state by visiting the USDA National Institute of Food and Agriculture's Family & Small Farms website at www.nifa.usda.gov/familysmallfarms.cfm for contact information and links to all state programs, or search for "Organic Farming Association," "Sustainable Farming Association," "Small Farm Institute," and "Small Farm Program" with a state name. One of the oldest and largest organizations is unfortunately not listed on the USDA site; this is the New England Small Farm Institute at www.smallfarm.org, which offers a plethora of resources as well as a mentoring program for new farmers.

Learning to Run a Business

Business planning programs with a focus on small-scale agriculture are run by universities and university-agricultural Extension services in a number of states, as well as by some independent sustainable ag groups, and are often free or low-cost. Begin by checking the www.start2farm.gov website, and consult your state university and county Extension agent to see what they have to offer, especially for financial and business planning assistance and training. You might also visit Americas Small Business

Development Center Network website at asbdc-us.org to find government-sponsored small business centers in your area. The organization's purpose is to help new businesses; though not focused on agriculture, there are still useful resources here.

PREPARING FOR START-UP

At some point in your farming career trajectory you will have had enough training and be ready to start your own operation. At this critical juncture a farm incubator program — that will give you some land to work and resources to go with it — or an experienced farm mentor who will help guide you through those first few seasons, can make all the difference.

Mentoring and Farm Incubator Programs

Mentoring/farm incubator type programs typically are designed to move a new farmer with some farm experience and business acumen into the start-up phase of a farm operation. They provide a terrific opportunity for a new farmer who is prepared but has few resources. Most will require that you meet criteria to be admitted; hence the need for having some experience, as discussed above.

In general you'll find that this type of program offers a nice mix of technical and financial assistance, mentoring from experienced farmers, help in locating land, and sometimes (though not often) help in financing a land purchase or lease. Some websites and programs are listed below as examples of what is available; be sure to look for additional opportunities in your area through your state and regional organic and sustainable agriculture associations, and university Extension.

ATTRA-NCAT. Lists mentoring/incubator-type programs on its website at www.attra.ncat.org.

The Land Stewardship Project — Farm Beginnings. This program's combination of classwork and hands-on mentoring from experienced farmers has been so successful in Minnesota that it has spread to several other states. The original program's website at www.farmbeginnings.org has links to all the other programs.

ONCE YOU GET YOUR LAND, keep learning! Sustainable agriculture field days are excellent places to learn practical skills and network with other farmers.

SCORE (formerly the Service Corps of Retired Executives). Provides mentoring by experienced business professionals for new businesses. Some of these executives have or are currently working in ag-related industries or have on-farm experience; you should get some excellent and (in my experience) cheerful and practical individually tailored advice on running your farm business.

Sustainable Ag Education

If for any reason you are not ready for that level of mentoring or incubating, another alternative is to enroll in a sustainable/organic ag education course or degree program which would give you the credentials to apply for higher-level jobs in the sector.

The National Center for Appropriate Technology's National Sustainable Agriculture Information Service at www.attra.ncat.org has a lengthy list of college and university courses and degree programs in sustainable ag and related fields.

The Midwest Organic and Sustainable Education Service (MOSES) at www.mosesorganic.org has a list of university and college programs. Though it is focused on the Midwest, some programs from other regions are listed as well.

OTHER WAYS TO GET ON LAND

ALTHOUGH ACQUIRING LAND can be difficult, persistence, creativity, and an open mind about how land tenure is structured should get you on a farm in the end. Listed below are some strategies for getting onto some land without having to come up with a down payment, or at least not the entire sum, on your own.

Leasing

Leasing land is an excellent and quite common way of getting a farm business going since you don't have to come up with a cash down payment. Don't lease land unless water comes with it; if the water is located at some distance from the land, you'll have to invest in some tanks to haul it.

A year-to-year lease is okay if you're just trying things out, but a multiyear lease is usually preferable, especially if you will be investing in infrastructure such as a greenhouse, irrigation system, or fencing. You will also need a multiyear lease if you intend to farm organically and the land has been farmed conventionally, since it will take three years to become certifiable in most cases.

Check Craigslist and other regional want ads, put the word out in the neighborhood that you're looking, and don't be afraid to knock on some doors and politely ask about renting.

Farm Link

Farm link programs connect new farmers with retiring farmers who want to help the next generation get started — and to find good future owners for their farms. Many states have programs, listed at www.farmtransition.org. In some cases, new farmers may be able to work themselves into eventual ownership of the farm; how each individual agreement is structured is between the owner and the new farmer.

Ganging Up

If you can gather a group of several new farmers, it may be possible to purchase land as a group and then divide it into separate parcels. Make sure that local ordinances allow this, and be absolutely certain to have a lawyer draw up the agreement among the various purchasers. One bad apple, or just someone changing their mind, can spoil the deal for everyone if you don't protect yourselves.

Or, if you're the only farmer but can assemble a group of people who want to support local food by buying land, you could set up a community-owned farm. The key here is to have an attorney knowledgeable about how to legally structure a business draw up an agreement to minimize the tax burden and to ensure that each shareholder would be entitled to their money back after a certain number of years if the farm fails, and to a return on their investment if it succeeds. If you can't think where you could find a bunch of investors, check out the Slow Money website at www.slowmoney.org for ideas, inspiration, and the essential connections for putting together this type of enterprise.

AND A FEW MORE IDEAS

Intentional Communities are all over the map, not only physically but in their activities, terms, and costs. Many are rural and some may suit you; check out the possibilities at www.ic.org.

SPIN ("Small Plot IN-tensive") Farming focuses on profitable vegetable production on less than an acre, including in backyards, community garden plots, and rented or borrowed land. The website at spinfarming.com provides a list of active SPIN farmers; contact one of these to see how the system is working for them and to become acquainted with this approach.

Shared Earth connects those who want to garden with those who'd like to see their land gardened — a logical place to start if you're interested in SPIN farming. Check out the listings at www.sharedearth.com.

The Farmer Veteran Coalition partners with numerous other organizations to offer workshops, farm tours, individual consultations — including on how to get access to land — and a help line for service members. If you've served in the military, check it out at www.farmvetco.org.

DON'T STOP LOOKING TILL YOU'RE HOME

Who knows what else is out there? Microfinancing, land trusts, living history farms in need of farmers? Just be absolutely certain, before you commit to any alternative methods of getting onto land, that it is legitimate and not a scam, and that you are legally protected.

And don't quit looking until you're home, on your very own farm.

GLOSSARY

Acre. 43,560 square feet; equivalent to 0.4 hectare. A perfectly square acre is 208.7 feet on a side. There are 640 acres in a square mile or "section," a common agricultural term. A half section is thus 320 acres, and a quarter section is 160 acres.

CAFO. Concentrated animal feeding operation, as in cattle feedlots, megadairies, confined pig and poultry (turkey, chicken) operations.

Certified Survey. A survey of property boundaries performed and recorded by a licensed surveyor.

CRP Land. Land enrolled in the USDA's Conservation Reserve Program. CRP pays farmers an annual rent on the entered acreage; in exchange, the farmer does not farm the land. This protects land from erosion and establishes wildlife habitat. CRP land has not normally been treated with synthetic chemicals and may be immediately eligible for organic certification.

CSA. Community-supported agriculture, the accepted term for a farm business structure where instead of direct sales of products, the farmer offers member shares in the season's farm production for a set price in advance. There are vegetable, meat, egg, dairy, and grain CSAs, and many other types and combinations.

Easement. A use granted by the owner of the land to another party. Common examples are allowing a landowner with no other access to his or her land to cross your property, and power line easements given to electric companies. If the property you are considering has access only across other private land, do not under any circumstances buy the property unless you have a legal, written easement across that other land.

Gray water. Wastewater from a house, excluding the "black water" from toilets. If you are looking at a farm with a gray-water system (such as emptying the wastewater into the road ditch), realize that even though the current owners are allowed to have it (since it's probably been in place before any wastewater disposal ordinances were passed) the new owner will be required to install a system that is up to code.

Infrastructure. Off-farm infrastructure includes roads and utilities (in rural areas, available utilities are typically electric and phone; municipal water is available only in some areas, and sewer is not generally available at all in areas of low-density population). On-farm infrastructure refers to buildings, fences, irrigation systems, and other permanent structures.

Perc test or percolation test. This test assesses the rate of water absorption and the amount of water soil can hold, in order to determine a site's suitability for a septic system.

Plat book. The county plat book, available in 12 states, is a map of all property boundaries in the county (except in subdivisions or towns). In the back is a list of full names and addresses.

RSA. Restaurant-supported agriculture. Similar to a CSA (see above), except members are restaurants rather than individuals.

Section. A square mile. A section contains 640 acres (259 hectares), so a half section is 320 acres (129 hectares), or 1 mile by ½ mile, and a quarter section is 160 acres (65 hectares), or ½ mile by ½ mile. A square 40-acre parcel (16 hectares) is a quarter mile on a side. A perfectly square acre is 208.7 feet on a side.

Slope (or grade). The degree of slope is expressed as a percentage, so that a 10 percent slope means the ground rises 10 feet (3 m) for every 100 feet (30 m) of horizontal distance.

ACKNOWLEDGMENTS

My warmest thanks to editor Deb Burns for suggesting this book and guiding it through to publication, and to all the great folks at Storey Publishing — Maryellen, Pam, Sarah, and the rest — who turn writers' efforts into such beautiful books, and are dedicated to supporting sustainable living in the midst of an unsustainable world. Many thanks also to Dan Masterpole, our County Land Conservationist, who has given me much education and insight on land use issues and government regulation in our years of both serving on our town's Plan Commission.

RECOMMENDED READING

Hansen, Ann Larkin. *The Organic Farming Manual.* Storey Publishing, 2010.
A comprehensive and practical introduction to small-scale organic farming, with special attention to finding the right land and suiting the farm enterprise to the farm, the markets, and the region.

Macher, Ron. *Making Your Small Farm Profitable.* Storey Publishing, 1999.
This crisply written book by the editor of *Small Farm Today* magazine is packed with excellent advice on small-scale farming and direct marketing, and includes some brief sections on assessing land and access to markets. Macher also discusses how to fit your enterprise to your land.

Scher, Les and Carol. *Finding & Buying Your Place in the Country,* 5th ed. Dearborn, 2000.
Good, pithy advice on rural land transactions.

Seltzer, Curtis. *How to Be a Dirt-Smart Buyer of Country Property.* Infinity Publishing, 2007.
At 750 pages, this book is massive, and full of discourses on tangential topics. It's also a gold mine of information on all aspects of buying and selling rural land for all sorts of purposes.

Stone, Lin. *How to Buy Land at Tax Sales: A Step-by-Step Guide to Buying Land at Tax Sales.* Truman Publishing, 1998.
An easy-to-read manual packed with specific information on government sales of tax-forfeit properties. If you choose to look into this type of real estate sale, this is an essential read since there are many technical details.

INDEX

OTHER STOREY TITLES YOU WILL ENJOY

Greenhorns, edited by Zoë Ida Bradbury, Severine von Tscharner Fleming, and Paula Manalo.
Fifty original essays written by a new generation of farmers.
256 pages. Paper. ISBN 978-1-60342-772-2.

A Landowner's Guide to Managing Your Woods, by Ann Larkin Hansen, Mike Severson & Dennis L. Waterman.
How to maintain a small acreage for long-term health, biodiversity, and high-quality timber production.
304 pages. Paper. ISBN 978-1-60342-800-2.

Making Your Small Farm Profitable, by Ron Macher.
An indispensable guide, full of advice on planning, marketing, and farming, to make your small farm equal big pay.
288 pages. Paper. ISBN 978-1-58017-161-8.

The Organic Farming Manual, by Ann Larkin Hansen.
A comprehensive guide to starting and running, or transitioning to, a certified organic farm.
448 pages. Paper. ISBN 978-1-60342-479-0.

Starting & Running Your Own Small Farm Business, by Sarah Beth Aubrey.
A business-savvy reference that covers everything from writing a business plan and applying for loans to marketing your farm-fresh goods.
176 pages. Paper. ISBN 978-1-58017-697-2.

Successful Small-Scale Farming, by Karl Schwenke.
An inspiring handbook to introduce the small-farm owner to the real potential and difficult realities of living off the land.
144 pages. Paper. ISBN 978-0-88266-642-6.
